ANTARCTICA
WHAT EVERYONE NEEDS TO KNOW®

ANTARCTICA

WHAT EVERYONE NEEDS TO KNOW®

DAVID DAY

OXFORD
UNIVERSITY PRESS

OXFORD

UNIVERSITY PRESS

Oxford University Press is a department of the University of Oxford. It furthers the University's objective of excellence in research, scholarship, and education by publishing worldwide. Oxford is a registered trade mark of Oxford University Press in the UK and certain other countries.

"What Everyone Needs to Know" is a registered trademark of Oxford University Press.

Published in the United States of America by Oxford University Press 198 Madison Avenue, New York, NY 10016, United States of America.

Library of Congress Cataloging-in-Publication Data
Names: Day, David, 1949– author.
Title: Antarctica / David Day.
Description: New York, NY : Oxford University Press, 2019. |
Series: What Everyone Needs to Know
Identifiers: LCCN 2019004677| ISBN 9780190641313 (paperback) |
ISBN 9780190641320 (hardback)
Subjects: LCSH: Antarctica–History. | BISAC: HISTORY / Polar Regions. |
SCIENCE / Earth Sciences / Geography.
Classification: LCC G870 .D295 2019 | DDC 919.89—dc23
LC record available at https://lccn.loc.gov/2019004677

1 3 5 7 9 8 6 4 2

Paperback printed by Sheridan Books, Inc., United States of America
Hardback printed by Bridgeport National Bindery, Inc., United States of America

CONTENTS

2 The Race for Antarctica in the Twentieth Century 33

3 Imperial Rivalry 67

4   War on the Ice   84

5   Science and Discovery   103

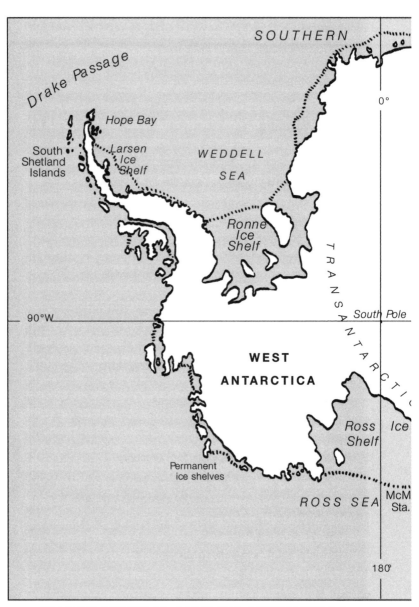

GEOGRAPHICAL MAP OF ANTARCTICA

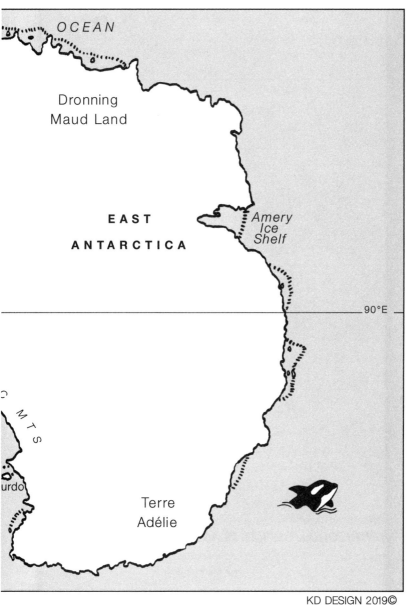

OCEAN

Dronning
Maud Land

EAST

ANTARCTICA

Amery
Ice
Shelf

90°E

MTS

urdo

Terre
Adélie

KD DESIGN 2019©

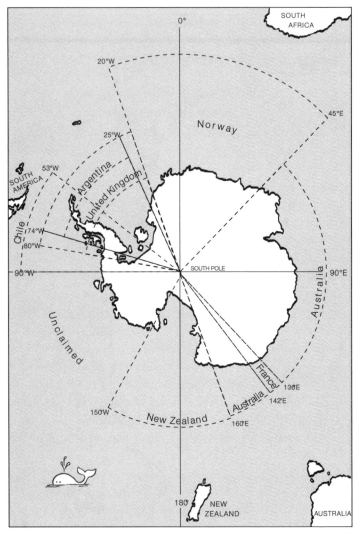

TERRITORIAL CLAIMS IN ANTARCTICA KD DESIGN 2019 ©

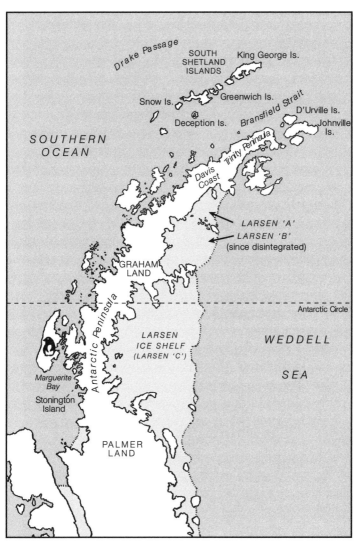

ANTARCTIC PENINSULA

KD DESIGN 2019 ©

ANTARCTICA

WHAT EVERYONE NEEDS TO KNOW®

OVERVIEW

Antarctica is the driest, coldest, and windiest continent on Earth. It is also the most inhospitable, being almost totally covered by an ice sheet that is up to 5 kilometers (3 miles) deep. This ice sheet contains 90 percent of the world's fresh water and largely conceals the mountain chains that cross the continent. At just over 13 million square kilometers (5.1 million square miles), Antarctica is the world's fifth-largest continent. It has no permanent inhabitants, and most of its interior has no life of any kind. Yet an amazing abundance of life can be found around its coastline and on its offshore islands—from massive blue whales to seals, birds, and penguins, right down to the tiniest life forms upon which all these larger animals depend. We are beginning to realize that the future of humanity is also inextricably bound up with what happens in Antarctica, as our burning of fossil fuels exacerbates the process of global warming and accelerates the melting of the Antarctic ice sheet. West Antarctica, including the Antarctic Peninsula, is warming faster than any other place on Earth, and the larger and colder region of East Antarctica now also shows signs of warming. Thus the need for action is urgent.

For millennia, Antarctica was beyond the reach of humans. Although ancient Greek myths suggested that there must be a continent at the South Pole, it took a mighty effort by European explorers in the early nineteenth century to confirm that supposition. Their first sightings seemed to offer nothing of value to those who sailed close to its forbidding coastline. The lucrative trade in the products of seals and whales—fur and oil—soon changed that assessment. This trade ignited a race to exploit Antarctica's marine resources and even to lay claim to the territory itself. Under the stress of that exploitation, the populations of seals and whales came close to extinction. While most of Antarctica's marine resources are now protected, the prolonged competition to own the continent continues. Despite the Antarctic Treaty of 1959, which was meant to end the territorial rivalry, nations still vie for control. Over the last half-century, they have used scientific bases and other symbolic means to lay claim to particular parts of the continent or even to the continent as a whole.

There are many reasons why Antarctica is coveted. Commanding the southern reaches of the world's oceans, Antarctica has obvious strategic importance for world powers as well as for the nations that lie close to it, particularly Chile and Argentina. For decades, oil companies have aspired to exploit the reserves of oil that lie beneath the seas of Antarctica, while trawlers have searched its ice-flecked seas for the rich bounty offered by the delectable toothfish and swarms of krill that sustain much of the Southern Ocean's marine life. Explorers and adventurers have also looked to Antarctica as the ultimate arena in which to test their stamina and make their reputations. However, the real value for humanity is to be found in the scientific discoveries that are being made beneath Antarctica's ice cap and offshore waters.

While Antarctica has gradually come within the compass of the world, the investigation of climate change has given the continent a global importance that would have been beyond the imagination of its early explorers. The warming of the surrounding seas and atmosphere has accelerated the glacial melt that is causing rising sea levels, while increasing absorption of carbon dioxide by the Southern Ocean threatens the survival of its marine life. Uncovering the secrets of Antarctica has become more critical than ever.

1

FIRST CONTACT

*How did Antarctica become the desolate, ice-cloaked continent
we know today?*

Hundreds of millions of years ago, Antarctica was part of a much
bigger continent. Dubbed "Gondwana" or "Gondwanaland"
by the Austrian geologist Eduard Suess, different parts of the
continent broke away and gradually moved north to form the
other continents of the southern hemisphere. The movement
occurred over tens of millions of years, beginning with just a
few millimeters a year as each land mass slowly separated.

Around 30 million years ago, after Australia had drifted
away, Antarctica became a separate continent. But Antarctica
was a very different continent and world than it is today. It
was a much more temperate place, with dinosaurs grazing
in grasslands or hunting for prey in woodlands. Fossil
discoveries have shown that small animals, like the marsupials
of present-day Australia and South America, were also found
in Antarctica, along with trees and other plants that still thrive
in other continents but have long been absent from Antarctica.

The disappearance of animals and plants was caused by
a succession of ice ages. In a process that began 1–2 million
years ago, the Earth cooled to such an extent that no animals or
plants could survive in the interior of the Antarctic continent.
Today, it is only on the coast that some species of penguins,

birds, and seals can survive, while much smaller life forms can be found on the few areas of exposed rock or in the waters of several unfrozen lakes, including those that have been discovered beneath the ice cap.

Although mountain ranges straddle Antarctica, most of the mountains are hidden beneath the ice cap. While the bedrock of the South Pole is only 30 meters above sea level, it is covered by an ice sheet that is nearly 5 kilometers (3 miles) deep in one place and averages more than 2 kilometers (1.2 miles) deep across the continent (nearly 3 kilometers of ice). This ice cap is the result of snow being deposited and compressed into ice over many millennia and inching inexorably toward the surrounding ocean before breaking off into gigantic icebergs. It was the impressive size of these icebergs that convinced the first sailors who ventured into the Southern Ocean that a large land mass must lie beyond their view. Indeed, some icebergs were so big that they were mistaken for the unseen continent itself.

Although it once had a temperate climate, Antarctica has never had any permanent human inhabitants. For as long as humans have been on Earth, the Antarctic environment has been so severe and dangerous that it has discouraged any thought of people living out their lives there. It is the coldest, windiest, and driest continent on Earth, with katabatic (downslope) winds often reaching hurricane force as they sweep down from the ice cap toward the coast. Antarctica has been kept in its frozen state by the Antarctic Circumpolar Current that continually circles the continent in a clockwise direction. The current is driven by prevailing westerly winds that cause the freezing waters to block any warmer surface waters from moderating Antarctica's climate. It was the action of this current that originally cemented in place the continent's ice cap, keeping the continent in its frozen state and minimizing the effects of global warming. That is fortunate for humanity, since 90 percent of the world's fresh water is locked within Antarctica's icy embrace. Unfortunately, there are signs that even Antarctica cannot remain immune to the effects of global warming.

How was Antarctica depicted prior to its discovery?

The ancient Greeks rarely strayed beyond the lands around the Mediterranean, but that didn't prevent them from having ideas about what might lie beyond the ends of their world. Aristotle argued that there must be a massive undiscovered continent far to the south. Otherwise, he said, the weight of the continents in the northern hemisphere would prevent the Earth from maintaining its regular movement through space. The Greeks could not imagine a world that was composed largely of oceans rather than continents.

The geographer Claudius Ptolemy from the city of Alexandria agreed with Aristotle's view and proceeded to draw a map of the known world based upon existing Roman knowledge. Ptolemy wanted to use all the sources that were available in the Great Library of Alexandria to concoct a complete world map. It included a massive continent dominating the southern hemisphere. Ptolemy's depiction of a "Great South Land" that was much bigger than Antarctica became a generally accepted fact, one that would take centuries to refute.

From the late 1400s onward, as printing developed, European geographers created and disseminated vivid representations of Ptolemy's imagined land on their maps and globes. These depictions were regarded as so authoritative that even Christopher Columbus took a copy of the map on his voyage across the Atlantic in 1492. Although geographers often called the land *Terra Australis Nondum Cognita* (Latin for "the southern land not yet known"), their maps included images of fantastical animals. The center of the supposed continent was the South Pole. Curiously, the continent was usually depicted by these imaginative mapmakers as being free of ice.

Since their representation was meant to balance the great continents of the northern hemisphere, the mapmakers showed the southern continent as enormous, often stretching north until it almost reached the equator. In their imagination, the continent was suitable for human settlement and abounded in valuable resources. Whether people already lived there had to remain a matter of conjecture until explorers were able to venture that far south.

Why did nations aspire to discover the South Pole?

Once Spanish explorers began returning from the Americas in the sixteenth century with their ships laden with gold and silver, there was a rush by rival European nations to find other undiscovered lands. They particularly wanted to find the mysterious Great South Land that loomed so large on medieval maps. If the Spanish and Portuguese could reap such riches from Central and South America, as well as from the newly discovered lands of the Indian and Pacific Oceans, surely there must be similar riches to be won from the continent in the southern hemisphere. It would take centuries before these assumptions were shown to be mistaken.

European explorers gradually began to realize from their voyages in the seventeenth and eighteenth centuries that the Great South Land could not be as great as the early geographers had portrayed it. As they criss-crossed the oceans, these explorers encountered mostly water and a few scattered islands. In the 1700s, the Dutch and Portuguese blundered into the north and west coasts of what they imagined must be the southern continent. But further exploration by the Dutch explorer Abel Tasman in 1642 and the English explorer James Cook in 1770 revealed that it was an island continent and much smaller than the mapmakers imagined. At first sight, it also seemed to lack the riches that other continents had in abundance. It was variously named "New Holland" or "Terra Australis," and later the name was changed by the British to "Australia." Many were convinced that there must be another, larger continent waiting to be discovered, one that would likely stretch all the way to the South Pole and have the riches that appeared to be plainly absent in Australia.

It wasn't just gold and silver that Europeans wanted. There was also wealth to be found in spices as well as in the furs of the sea otters and seals that Chinese traders were anxious to exchange for the tea that had become so popular in Europe. As trade shifted eastward into the Pacific and Indian Oceans, the discovery of a southern continent could also have great strategic significance for any empire that wanted to control the lands that abutted those distant oceans. Explorers and sealers searched further and further south for the land they were certain must be there. But it wasn't until the early nineteenth century that they finally found it and realized how different it was from to the imagined land depicted on medieval maps and globes.

Who first discovered Antarctica?

Six of the seven continents were peopled in the centuries after modern humans emerged from Africa at least 70,000 years ago. Since there is no written record of their discovery, we cannot know the identity of the first people to walk on those continents or to land on their shores. The last continent to be found was Antarctica. Even though it was discovered in modern times, the identity of its discoverer remains a matter of some dispute.

The Maori people of New Zealand believe one of their ancestors sailed so far south that he encountered the ice cliffs of Antarctica. The story of such a voyage is part of their oral history. Over the centuries, Polynesian people had indeed made many long voyages across the vast distances of the Pacific, during which far-flung islands were discovered and settled. An enterprising Maori mariner might well have ventured south one summer, guided by the stars, before returning with stories of an ice-covered land. However, it is impossible to know for certain whether an adventurous Maori explorer was the first to sight Antarctica, or whether he was misled by a cloud bank or merely saw a large iceberg that had drifted north toward New Zealand.

European ships were more equipped to face the rough waters of the Southern Ocean, and Europeans usually kept scrupulous written records of their voyages. Their logbooks should have allowed the name of the continent's first discoverer to be beyond doubt. However, American and British whalers and sealers were often secretive about their travels for fear of letting rivals know about rich sealing grounds or waters that were abundant in whales. Although it is possible that one of their vessels may have ventured sufficiently south in the late seventeenth or early eighteenth century to see the Antarctic continent, no written record of such a sighting has so far surfaced.

This brings us to Captain James Cook, who sailed three times into the Southern Ocean looking for the Great South Land. On his second voyage, from 1772 to 1775, Cook came close to finding it but pack ice prevented him. Although he claimed an island off the Antarctic Peninsula for England, mistakenly believing that it was part of the fabled southern continent, he quickly realized his mistake. He consoled himself with the thought that the elusive continent was not worth the trouble of discovering, arguing that its harsh climate and icy defenses would make it a useless acquisition.

It was nearly half a century later that the continent was finally sighted. The identity of its discoverer remains a matter of some contention. There are three claimants for the honor. One was an American sealer, Nathaniel Palmer, whose search for seals took him to the islands off the tip of South America. The second was a British sealer, William Smith, who had spotted some of these same islands and believed them to be continental in size. Excited by the news, the Royal Navy sent him back to confirm his discovery, with his ship placed under the command of naval officer Edward Bransfield. The third claimant was the Prussian-born explorer Captain Gottlieb von

Bellingshausen, who was sent south in 1819 by Tsar Alexander I as part of the Russian Empire's expansion into the Pacific.

It is extraordinary that all three explorers happened to converge on Antarctica in 1819–1820, with their voyages allowing each to claim they had been the first to discover Antarctica. Bransfield and Smith caught sight of the Antarctic Peninsula on January 30, 1820, and thought it could be part of a continent. The sea conditions were too dangerous to find out for sure, and they made no attempt to step ashore and claim it for Britain as they had done on some islands discovered earlier. Nevertheless, their sighting allowed British historians to claim that Bransfield should be credited with the discovery of the continent.

American historians have been just as vociferous in supporting Palmer, who may have seen the Antarctic Peninsula from a high point on nearby Deception Island and even sailed there to check out whether there were seals on its shores. Although an American writer would later publish an account that purported to describe a meeting on Deception Island between Palmer and Bellingshausen, during which Palmer supposedly told the Russian of having seen the continent, no documentary evidence has been found to support this claim. In particular, there is no mention of it in Palmer's logbook.

As for Bellingshausen, there is convincing evidence that the Russian explorer sighted the coast of Antarctica on January 28, 1820, just two days before Bransfield saw land on the other side of the continent. Moreover, accounts by Bellingshausen and his officers make clear that they were aware it was a continent rather than just a frozen sea. In a letter to the Russian Minister for the Navy, Bellingshausen described seeing "continuous ice, at whose edges were pieces piled one upon another, and with ice mountains seen at different places in a southerly direction."[1] Nonetheless, debates about the real discoverer of the continent endured for more than a century.

Why is the debate as to who discovered Antarctica still relevant?

Although as noted the historical evidence points toward Bellingshausen as having been the first to discover Antarctica, the Russian claim was long disputed. Bellingshausen did not help his own cause by being tardy in publishing his account of the historic voyage. A Russian edition of his journal did not appear until ten years after the voyage. His recognition was further delayed by the absence of an English edition until 1945. When the account was published in English, the translation was so poor that it created considerable doubt as to whether Bellingshausen had seen the actual continent or merely glimpsed an ice shelf that was perhaps not even attached to Antarctica.

The assumption that it was either Palmer or Bransfield who should be honored as the discoverer of Antarctica endured for than a century. The issue was of more than just academic interest since a definitive answer would reinforce the respective British and American territorial claims over Antarctica. From the 1920s onward, both the U.S. State Department and the British Foreign Office sponsored mapping work and historical research that was designed to support the claims of Palmer and Bransfield respectively. The American and British governments were trying to buttress the territorial claims being made by their current explorers, and they accused each other of fabricating evidence.

Amid the ongoing furor, the Russian claim was disregarded due to the relative lack of supporting evidence. It was only in 1951, after a newly translated account of his voyage was published along with other supporting documentation, that Bellingshausen was gradually acknowledged by most British and American historians as being the first. The question would become less important once the Antarctic Treaty of 1959 set aside territorial claims that had been vigorously prosecuted for several decades. Even now, a few historians continue to doubt Bellingshausen's claim, suggesting that he had only seen an ice shelf without realizing it was of continental extent. While the argument about who was first to sight Antarctica might persist, there is less argument about who was first to step ashore onto its coastline.

Who was the first to actually step out onto the Antarctic continent?

The honor of being the first ashore did not go to an explorer but to an American sealer who was looking for new seal colonies to exploit after those on nearby islands had been wiped out. In February 1821 Captain John Davis of the schooner *Celia* cautiously threaded his way through icebergs and tricky currents toward the land that had been sighted the previous year by Bransfield and named as "Trinity Land." It was part of what we now know as the Antarctic Peninsula. Today it has become one of the most keenly contested lands on Earth though it didn't seem worth contesting when Davis saw it. He was hoping to find seals, but there were no seal colonies on the land where his crew could make a quick killing.

After the crew had gone ashore and returned—through an enveloping snowstorm—with nothing in their boat, Davis had to be content with noting in his logbook that he had seen "a Large Body of Land" that was "high and covered entirely with snow."[2] Davis had seen the offshore islands and sailed around them, but this sighting seemed different. He could see no end to it and wrote, "I think this Southern Land to be a Continent." Of course, he had no way of knowing from his brief foray that he had landed on a tongue of land that was joined to a continent covering the South Pole and stretching across the Southern Ocean for thousands of miles. All Davis knew was that it seemed larger than the string of nearby islands. His feat was slow to be recognized by the world. It was not until the 1950s that an American historian chanced upon the logbook of the *Celia* and wrote about their role in the discovery of the continent.

How did the earliest explorers react to Antarctica?

Sailing south until they encountered massive icebergs, some eighteenth-century sailors, such as Cook, were led to think that the long-searched-for continent was within sight. The combination of sea mist and white clouds ballooning above the ice pack gave the impression of mountains on the far horizon. One of the scientific members of Cook's second voyage in 1772 reported the excitement when an officer clambered down from the masthead early one morning, claiming to have seen land in the distance. On hearing the commotion, all aboard came up from below deck to see a vast field of ice with what they took to be mountains in the distance. But there was no land there, as Cook himself discovered when he sailed over the same route three years later and found there wasn't even any ice. The sailors had simply molded clouds into mountains. They had seen what they expected to be there.

The Great South Land of the European imagination was also supposed to be peopled by members of a civilized society. Cook's three voyages, as well as those of other explorers, showed that any land to be found was situated much further south than medieval mapmakers had supposed. Consequently, there was little likelihood of finding any inhabitants, and certainly not the sort of civilized societies that the Spanish and Portuguese had encountered in the Americas. Instead, there were penguins, seals, and whales along with a multitude of birds. Mapmakers had suggested that more fantastical animals would be found, but these were extraordinary enough. Specimens of those animals that could be caught or killed were collected for the entertainment and wonder of European audiences or merely to provide a welcome change from barrels of salted meat for the sailors. As a crewman on one of Cook's ships recorded in 1775, "several of the gentlemen in both ships diverted themselves in shooting penguins, and the sailors had no less pleasure in eating them, than the gentlemen had in killing them."[3]

The earliest explorers had a tendency to invest penguins with human characteristics. Indeed, when seen standing upright on a distant ice floe, their shape and coloring meant that they were sometimes mistaken for humans. It is not surprising that the largest of the penguins was called the "emperor penguin," while a slightly smaller species was called the "king penguin." There was even a sense of their being the original and rightful occupiers of the place. When the French explorer Jules Dumont d'Urville landed on a rocky islet off the Antarctic coast and proceeded to claim the region for France, the occupying penguins were unceremoniously shoved aside to make way for the new owners.

The Antarctic icebergs lent some credence to these anthropomorphic fantasies. The early explorers had expected to find a previously unknown civilization, as Columbus had done in the Caribbean, and the shapes of the icebergs appeared to resemble the ruins of some great empire. Dumont d'Urville had explored the remains of ancient Mediterranean empires and likened the icebergs to a great city reduced to ruins by an earthquake. Indeed, it was common for sailors to think they were looking upon the remains of human constructions. These men were not only the first explorers of Antarctica but also imagined themselves the conquerors of it.

What was the nature of the Southern Ocean sealing trade?

The explorers certainly were conquerors of the animal population. Although Cook had disparaged the idea of Antarctica being of any value, his voyages of discovery lit a fuse that exploded into a burst of activity. When mariners read his reports of numerous whales and seals off the tip of South America, vessels were soon sent on speculative voyages to see what could be gleaned from those tumultuous seas. Sealers and whalers were not concerned with claiming the lands but with exploitation of their marine life. It was all about whales and seals. Although penguins had been seen in vast numbers, there was no money to be made from them, other than having several stuffed as curiosities for museums and collectors or having them butchered as food.

Most of the whale species that frequented Antarctic waters proved difficult to kill and haul in with methods used in the eighteenth century. Whalers in rowboats were no match for the fast-swimming, deep-diving humpback and blue whales and other similar species, which tended to sink to the bottom when they were killed. And there were still plenty of the slower-moving bowhead whales to be caught in the northern hemisphere.

The seals were a much easier proposition. The large colonies of elephant and fur seals had lived undisturbed on the rocky shores of the remote sub-Antarctic islands. Crowded together, and relatively defenseless, the lumbering elephant seals were easy to kill with guns, swords, or clubs. The males grew up to 6 meters or nearly 20 feet long and weighed up to 3,000 kilograms (over 3 tons). Thick blubber was cut from their massive carcasses and boiled down in cast-iron try pots until the oil could be poured off into barrels. The oil had a multitude of uses, from lighting to lubrication.

Fur seals were more agile, but they were in much larger numbers and proved just as easy to slaughter. They weren't rich in blubber but their skins were highly valued for making hats and coats. If sailors were prepared to face the dangers, there was a fortune waiting to be made by those who were willing to venture so far south. The English were the first to do so, but they were soon followed by American whalers and sealers from New England and by a few enterprising Argentinians. The ships returned with their holds filled with barrels of oil and salted skins that found a ready market on the wharves of New York and London.

South Georgia was one of the major centers of their activity. Set in the South Atlantic some 1,500 kilometers (900 miles) northeast of the Antarctic Peninsula, the large uninhabited island had been discovered by Cook in 1775. Just over

160 kilometers (75 miles) long, its untouched state was soon destroyed by the arrival of hundreds and then thousands of sealers who established their camps on its shores. Their massive try pots were kept boiling on the beaches with the blubber of elephant seals, while the surf ran red with the blood from the hapless creatures and the carcasses of the skinned fur seals. By the early 1820s, both fur seals and elephant seals were close to extinction, and there were no fortunes left to be made on these sub-Antarctic islands. It remained to be seen whether there were seal colonies yet to be found on lands further south.

Why were the early sealers unable or unwilling to create a sustainable industry?

The sealing trade could have kept hundreds of American and British ships and thousands of sailors in work and also prevented the almost complete annihilation of their quarry. However, that was never a realistic possibility. It has been estimated that on the island of South Georgia alone, 250,000 seals were slaughtered by about 1820. By the end of the decade, some 7 million sealskins had reached American ports. It would take many decades for the seal colonies to recover, not just on South Georgia but on all the islands of the far south that were chanced upon by those sailing the oceans in the wake of Cook and the whalers.

It might have been different had there been some oversight of the industry, but the sealers plied their trade far beyond the reach of governments. Anyway, they believed that the world was theirs to be plundered. After all, hadn't the Bible told them to "be fruitful and multiply, and replenish the earth, and subdue it: and have dominion over the fish of the sea, and over the fowl of the air, and over every living thing that moveth upon the earth"?

Practically speaking, no agreement to limit the killing could ever have been enforced in the early nineteenth century when the rewards were so great. And nobody was going to exploit seals sustainably when they knew that everyone else was going to kill without regard in the following season. Instead of perceiving the seals as a finite resource whose killing had to be managed, they treated a newly discovered seal colony like a gold mine to be emptied of its riches before someone else took it from them. Once the slaughter was done, the search for a new colony could begin. For a time there seemed to be no end to new seal colonies.

When the beaches and islands of South America were cleared of seals, the search went further afield to the islands of the Pacific, to Africa, and to the south. Although an increasing number of sealing vessels returned to their ports with their holds only partially filled, sufficient new seal colonies were found, keeping the hopes of sailors and their corporate supporters alive. And there was still much to discover in the furthest reaches of the Southern Ocean.

What was the connection between sealing and territorial aspirations?

The English were the first to make territorial claims in the Antarctic. Although Cook did not land upon the continent or even sight it, he did lay claim to several nearby islands and named them in honor of King George III, as well as his patrons at the Admiralty and in Parliament. With much ceremony, Cook would raise the English flag and fire a volley of shots on lands that he thought might comprise part of the Southern Continent. He was always disappointed to discover that the supposed continent was merely an island, albeit sometimes an extensive one. He concluded his search for the Southern Continent with the consoling thought that its ice-cloaked location would make it not worth owning. The seals, along with the imperial swing from the Atlantic to the Pacific, changed all that.

The seas off the tip of South America offered a strategic passage between the Atlantic and Pacific. Whichever nation controlled those seas would enjoy greater control over the Pacific and the growing commerce that passed between the two oceans. By the early nineteenth century, the British, Spanish, French, Russians, and Americans all harbored some desire to dominate the Pacific. And the wealth to be made from whales and seals gave greater force to the arguments of imperial propagandists who wanted their nations to lay claim to the South Pole.

It would have been easier for Britain and America had their sealers and whalers been inclined to establish permanent settlements at the places where they landed. But their camps tended to last no longer than a summer. When the supply of seals was exhausted, they would leave for another killing field, with the signs of their fleeting presence being gradually lost to the ravages of weather. Although an occasional British sealer would go to the bother of formally making a territorial claim when they encountered a new island, American sealers showed no such interest.

Most sealers who sailed among the islands off the Antarctic Peninsula, and occasionally hunted seals along other parts of the Antarctic coast, did leave evidence of their presence in the form of the maps they made and the names they might bestow on certain geographical features, such as Bransfield calling an island "New South Britain." However, that wasn't a sufficient basis for a nation to make a territorial claim, which would require an officially sanctioned expedition with authority to conduct a formal claiming ceremony, preferably on the lands of which they purported to take possession. Once the sealing trade declined in the early 1820s, the rationale for making territorial claims became less compelling from a commercial point of view.

Why was the continent itself left largely untouched and unexplored for so long?

The further south the sealers ventured in the late eighteenth and early nineteenth centuries, the more difficult and dangerous it became. There was the possibility of their ship being crushed by the ice pack or being caught fast in a quickly freezing sea. Either would mean almost certain death. There were dangers, too, from unknown currents, hidden shoals, shifting icebergs, and fog-shrouded coastlines. As if that was not enough to negotiate, captains had to deal with the ever-present threat of crews that could turn mutinous if the conditions became unbearable or the catch was insufficient to ensure a good payoff at the end of the voyage. These considerations meant that captains were cautious about venturing too far into the unknown.

One who did go further south than his counterparts was James Weddell, captain of a sealing vessel, the *Jane*, which was sent by its owners in 1822 to search for new sealing grounds. Weddell headed deep into the South Atlantic, steering the *Jane* to the east of the South Shetlands, believing from Cook's report that he would find the Great South Land there. But there was only open sea. He pressed on, going 320 kilometers (about 200 miles) further south than Cook had done. He could have gone even further into what would later become known as the Weddell Sea, as there was no ice that year to impede his passage. However, he could not see any coast on which seals might be found, and his sailors were becoming restless at the likely lack of profit. It would be another century before the ice conditions would allow Weddell's voyage to be replicated. Until then, all those who tried to follow his chart encountered impenetrable pack ice.

It might have been different had there been definite reports of multiple seal colonies on the shores of the Antarctic continent. However, the few sealers and explorers who sighted various parts of the coastline confirmed Cook's conclusion and returned with reports of a land largely devoid of exploitable life. Much of the coast was surrounded by ice pack or hemmed in by impossibly tall cliffs of ice that denied access to both humans and the animals they sought. There were indeed few ice-free shores to be found around the entire coastline of Antarctica on which seals could establish their colonies. As a result, the continent largely escaped the attention of man for another century.

Why did it take so long for whaling to begin in earnest in Antarctica?

Whalers had been scouring the oceans of the northern hemisphere for centuries, looking for the tell-tale blow of a whale pod as the animals surfaced and expelled air before diving down again into the depths or as their fins curved across the surface. Sperm whales have been recorded diving down to 3,000 meters (about 10,000 feet) and can remain below the surface for more than two hours. As the whalers' methods became more sophisticated by the end of the eighteenth century, and the demand for whale oil increased, the whales became more difficult to find in the North Atlantic. Undaunted, the whalers of England and those of the newly independent United States began to range further afield, seeking their quarry in the Pacific, where the pleasant climate and often welcoming islanders compensated for much longer voyages.

Migrating whales would leave their summer feeding grounds in the Antarctic to breed over winter in the mid-Pacific, which is where many whalers would search them out. It wasn't unusual for whaling ships to circumnavigate the globe, taking advantage of the prevailing winds in the southern hemisphere to speed their passage. The winds were stronger and more constant the further south they went. Once Australia was claimed by the English in 1788, the island continent became another center of whaling activity. Several whaling stations were established along the Australian east coast to prey upon whales migrating to the tropics and on those whales calving in its shallow bays, while whale-catching forays also extended to the sub-Antarctic islands off Australia and New Zealand. Sailors would get particularly excited when they came across the so-called right whales, which had baleens for filtering small prey rather than teeth. Right whales have a thicker layer of blubber than some other whales, which means they produce more oil and would float when killed rather than sinking to the depths. Apart from the oil, the flexible baleens were also valued for their use in the hooped dresses of fashionable women. As long as whales could be found in sufficient numbers in warmer waters, there was no incentive for whalers to chance their lives and their ships in the much more hazardous waters of the Antarctic.

The region was hazardous because most of it remained unseen and unmapped. This gave the Antarctic a sense of mystery, which fascinated people and led some to make all sorts of wild speculations as to what might lie at the bottom of the Earth. One of the wildest theories was proposed by the American John Cleves Symmes, who helped inspire the United States to send a great exploring expedition to the southern seas.

How did people come to believe the Hollow Earth theory?

Until well into the nineteenth century, the world still knew little about Antarctica. As noted earlier, a few points on its coast had been seen by the Russian explorer Bellingshausen and the British explorer Bransfield, and sealers like Palmer may have seen the Antarctic Peninsula. None of them had seen anything that would encourage others to risk their lives in search of the South Pole. The ice and snow ensured that there were no precious metals visible. It also ensured that there were few places for seals to establish their colonies. That did not prevent some people from speculating wildly about what might lie at the base of the globe.

Britain and its European rivals had all established overseas empires and were still acquiring new colonies. Of the major powers, the United States was almost alone in not having an empire, other than its steadily expanding continental one (mainly at the expense of the indigenous populations). While Americans mostly eschewed acquiring an overseas empire in the manner of their former English overlord, the Antarctic offered the possibility of gaining an overseas territory without the complicating factor of controlling a native population. It also offered the possibility of answering one of the last great geographical puzzles of the modern world: What exactly did lie at the North and South Poles?

There were all sorts of theories. Some thought it might be just a frozen ocean, while others argued that the ancient Greeks were right in postulating the existence of a Great Southern Continent. One American went much further by suggesting in 1818 that the North and South Poles were entrances, thousands of miles across, to the interior of the Earth. Symmes, a former army officer, claimed that the entrance at the North Pole explained the origins of Native Americans as well as the annual appearance of great herds of migrating caribou and other animals, which had also emerged from this supposedly lush and temperate interior. He even constructed a wooden model of his hollow world to convince American audiences of his vision.

Symmes wanted to organize an expedition by reindeer-drawn sleigh from Siberia to find the northern opening, but no one was willing to accompany him or to finance the journey. It might have ended there had he not caught the attention of the crusading journalist Jeremiah Reynolds, who helped Symmes reach a wider audience before gradually making the Hollow Earth theory his own. From 1825, the pair toured the towns and cities of America's Eastern seaboard, lecturing about the wonders that might exist at the poles. It was the South Pole, rather than the North, that particularly interested Reynolds, who proved much more convincing than Symmes and attracted considerable public support.

By 1828, having ditched Symmes and his theory, Reynolds called on Congress to support an expedition to the South Pole for the cause of science and commerce. He argued that it would be a great achievement for the new nation to discover what lay at the South Pole, bringing great profit to the nation from seals and other lucrative fur-bearing animals that must surely exist there. The prospect of profit proved persuasive to the congressmen and President John Quincy Adams.

The Americans were not the only ones interested in exploring the Antarctic. The British government sent a scientific expedition there in 1828 to take astronomical observations that might help calculate the precise dimensions of the Earth. Led by Captain Henry Foster as astronomer, the expedition not only took observations but also claimed possession of part of the Antarctic Peninsula in the name of King George IV. The news of the British expedition lent urgency to the American plan, which was approved by Congress in May 1828, only to be scuppered later that year by the election of the more skeptical President Andrew Jackson.

Undaunted, Reynolds pressed ahead, gaining support instead from Captain Edmund Fanning, a sealer based in the New England port of Stonington (in Connecticut). Fanning's support saw the expedition being called the "South Sea Fur Company and Exploring Expedition." As the name suggested, it would have more to do with sealing than exploring. Three sealing vessels set off in October 1829, with the experienced explorer Palmer as one of the captains. Although styled as a scientific expedition that would try to push as far south as possible, it was soon overtaken by the profit-making imperative and concentrated instead on finding seals around the southern coasts and islands of South America. No new discoveries were made, and only a paltry haul of sealskins was stowed away in barrels for the return journey. Once again, whatever lay at the South Pole remained undisturbed and Symmes's theory of it providing an entrance to the center of the Earth remained untested.

Why did the pace of exploration accelerate in the 1830s?

The failure of America's private expedition to get anywhere near the South Pole only increased Reynolds and Fanning's desire to do so, and to do it the next time with government support. Far better for taxpayers to foot the bill and bear the risk in the hope that any discoveries would boost American commerce and prestige. Sealers could safely exploit any discoveries after they had been made and mapped. With this in mind, Reynolds and Fanning presented Washington with a new plan.

While the Americans were considering their options, a British sealing company sent two ships under the command of former naval officer Captain John Biscoe to search for new sealing grounds. For a decade the trading company owned by the Enderby brothers of London had reaped a handsome harvest from trading in sealskins, only to find their supply dwindling due to overexploitation. Reports of possible new islands at the South Pole prompted them to send Biscoe south in 1830. Despite a two-and-a-half-year search, Biscoe barely filled a single barrel with salted sealskins. However, he did sight a mountainous, snow-covered land far to the south of southern Africa and chanced upon new islands off the coast of the Antarctic Peninsula. Biscoe gave all his discoveries British names and even stepped ashore on the Antarctic Peninsula to claim it for Britain, naming it "Graham Land" after Sir James Graham, Britain's First Lord of the Admiralty. He suggested in his log that it was part of a much larger continent. It was the first time that any part of Antarctica had been formally claimed. The United States was now on notice, particularly as Graham Land was known to the Americans as "Palmer Land," in honor of its supposed discoverer, Nathaniel Palmer. If Washington didn't respond with an expedition of its own, whatever lay at the South Pole was likely to become British by default.

The pressure for an American expedition intensified in early 1836, when Reynolds addressed a meeting of congressmen during which he sketched out a grand plan for an expedition focused upon discovery. "No spot of untrodden earth" should go unexplored, declared Reynolds, who held out the prospect of the Stars and Stripes being unfurled to fly over the South Pole. With both the British and the French dispatching official expeditions of their own, the United States would have to move quickly if it was going to proclaim America's right to rule the Southern and Pacific Oceans, where its hundreds of whaling ships were constantly on the hunt.

The tantalizing prospect of reaching and possessing the South Pole found much support in the American capital, at least initially. Although the expedition was approved, some congressmen began to fear that the expedition was being sent to prove Reynolds's crackpot Hollow Earth theory. In the face of their skepticism, the focus of the expedition was shifted to the more temperate climes of the Pacific and was put under the command of the well-connected surveyor, Lieutenant Charles Wilkes, who ensured that Reynolds would not accompany, let alone lead, this new expedition. Delays in fitting out the ships, now totaling six vessels in all, meant that the expedition did not leave the United States until August 1838. The delayed departure meant that Wilkes was following in the wake of the French explorer and naval officer Dumont d'Urville, who had already explored the South Pacific.

Like the American, the French expedition was meant to be focused mainly on the trading and imperial possibilities of the Pacific. But when Dumont d'Urville headed south in 1837, his mind was set on making a name for himself. His first attempt at reaching the South Pole by penetrating the ice pack of the Weddell Sea was unsuccessful. He tried again and again and was repelled by the ice each time. Undeterred, he set sail for Australia, determined to strike south from Hobart the following summer and attack the Antarctic from the opposite side.

Before leaving Hobart in January 1840, Dumont d'Urville learnt that the Wilkes expedition had reached Sydney and was presumably intent on also heading for Antarctica. This gave

added urgency to his voyage. In the Great Game of exploration, there is no honor in coming in second. Dumont d'Urville
was relieved to come in sight of the Antarctica coastline on the
evening of January 19, 1840, but he was disconcerted to find
an imposing ice cliff that stretched as far as he could see in
both directions. It kept him from going further south or even
from landing ashore. Dumont d'Urville had to be content with
landing on a rocky islet offshore, where his men dispossessed
the penguin inhabitants and raised the French Tricolour as a
sign of them being the new owners of the islet and the adjacent Antarctic coastline. Dumont d'Urville named the territory
Adélie Land in honor of his long-suffering wife.

It was not a moment too soon. On January 30, one of
Wilkes's ships appeared out of the Antarctic mist and raced
past, its sails fully set. No communication passed between the
rival ships. Several of Wilkes's vessels had been sailing in sight
of the Antarctic coastline, but the American officers were often
unsure as to what they were seeing. What looked like land frequently turned out to be low clouds or a large iceberg. The
various American logs make no claim of a definite sighting of
land until January 30, 1840, when Wilkes saw an elevated area
with mountains beyond and noted in his journal: "Antarctic
Land discovered beyond cavil."[4] He'd sailed along the coast
for 1,300 kilometers (about 800 miles) and had seen enough to
be sure it was of continental extent. Naming it "Antarctica,"
based on the ancient Greek word for the land at the opposite
end of the world to the Arctic, he set sail for Sydney where he
learned that Dumont d'Urville had arrived in Hobart and was
claiming to have seen the supposed continent on the evening
of January 19. With the American consul looking on approv-
ingly, Wilkes wrote a statement for the Sydney press, declaring
that his expedition had first seen the continent on the morning
of January 19, thereby trumping the Frenchman by half a day.

The British explorer James Clarke Ross was a latecomer to
the party. Having located the North Magnetic Pole in 1831,
he had been sent by the British Admiralty to determine the

position of its southern counterpart. The magnetic poles are located separately from the geographical poles and their positions are always slowly shifting with the ever-changing magnetic field of the Earth. Near the equator, the difference between the geographical and magnetic poles is inconsequential for captains steering their ships with the aid of a magnetic compass. However, the further they venture south or north from the equator, the more crucial it becomes to know the difference between the two poles so that they can accurately ascertain their position on a map. Many shipwrecks and much loss of life has been caused by captains not adjusting their position according to that vital difference in bearing between the two poles, which can result in ships being out by miles.

Ross arrived in Hobart while Dumont d'Urville was still there. He was aware that his American and French rivals had beaten him to that previously unseen stretch of Antarctic coastline south of Tasmania. They'd also made their own estimates of the South Magnetic Pole's position, which threatened to thwart Ross's ambition to be the first to locate both the North and South Magnetic Poles. However, Ross hoped to outdo his rivals by going further south and thereby getting a more accurate estimation of the pole's location, and also by exploring in an easterly direction rather than heading west as Wilkes and Dumont d'Urville had done. He believed it would be demeaning for a British explorer to follow in the tracks of an explorer from another nation. One of Enderby's sealers had reported sighting an open sea far to the south of New Zealand. So that is where Ross headed with his two sturdy ships.

Although Ross was intent on calculating the position of the South Magnetic Pole, he was also instructed to claim any new territory he might discover and to reaffirm the claims of Antarctic territories that had already been discovered and claimed by British explorers. Unlike Wilkes and Dumont d'Urville, his passage was unimpeded by an ice cliff and he found the entrance to the promised open sea. Landing on offshore islands, Ross claimed the adjacent stretches of continental coastline as he went further and further south into what would later be named after him as the "Ross Sea."

After reaching 78° S, which was much further south than any other explorer had managed to go, he encountered two volcanoes, which he named after his two ships *Erebus* and *Terror*. He had wanted to plant the British flag at the South Magnetic Pole, but that would have required an expedition across the ice and snow. He had to be content with going further south than anyone else, gaining a more precise location for the South Magnetic Pole, and confirming on the return voyage that the part of the Antarctic continent that Wilkes had claimed to have seen was actually open ocean.

What scientific questions did these early explorers answer?

The most obvious aim of the earliest explorers was a geographical one: to fill in the huge blank at the bottom of the world where the Great South Land was meant to lie. For more than a century, ships had sallied forth from Europe in search of the mysterious Southern Continent. Many thought they had found it, including the Spanish explorer Fernandez de Quiros, who encountered an island near New Guinea in the early 1600s and mistakenly thought it was a continent bigger than Europe, and the Dutch explorer Abel Tasman, who chanced upon the west coast of the New Zealand islands in 1642 and thought that it was part of the sought-after continent. Cook proved that New Zealand was not a continent when he circumnavigated both islands in 1770. Yet when he discovered various islands on his three voyages and thought for a time that they might be part of the Antarctic continent, he was similarly mistaken, albeit briefly.

Explorers hoped that the elusive continent would prove as bountiful as Columbus's West Indies had been. Because of the cold climate, explorers knew they were unlikely to find the sort of valuable plants that had been discovered in warmer parts of the world, whether it was tea and tobacco or cloves and cinnamon. But any botanical specimens they found could still be collected and brought back to Europe or America for classification. This in turn might help to answer some of the great scientific questions of the time. For instance, if some of the plants were the same as those found on other continents, it would suggest that the continents had once been linked and later separated according to the theory of continental drift first propounded in 1596 by the Dutch mapmaker Abraham Ortelius. However, the ice and snow that was encountered on those parts of the Antarctic continent that were discovered in the early 1800s suggested that there were no plants at all to be found on this frozen continent.

What they did find was fossils of plants. The expedition of Fanning and Reynolds in 1829 included a four-man scientific corps led by Dr. James Eights from Albany's Lyceum of Natural History. It was poorly equipped for the work that Eights had hoped to undertake, and it never got close to the Antarctic continent. But the vessels did visit the South Shetlands and other sub-Antarctic islands, where Eights gathered scientific specimens while the sailors tried to turn a profit by killing whatever seals they could catch. Although Eights was disappointed with his scientific haul, and the expedition was a financial failure, he did return with fossils from the South Shetlands that proved those treeless islands had once harbored lush forests.

The discovery of Antarctic animals might also have helped to answer the debate over continental drift. If animals that were present in Australia or South America were also found in the Antarctic, there surely must have been some sort of connection between them in the distant past. However, there were no land animals to be found on these islands and none to be seen on those parts of the Antarctic coast that had been scanned by explorers through their telescopes. Neither were there any of the valuable sea otters that had been found in abundance in the North Pacific. There were penguins aplenty, and some were killed, skinned, and stuffed for exhibition back home, but they were no longer such a curiosity by the late 1820s.

Collecting rocks from Antarctica might also have answered questions about the origins of the continents, but the rocks gathered by Eights and others came from the islands rather than from the continent. Moreover, explorers were more intent on discovering gold and silver, and there were no precious metals to be found. It might have been the Age of Scientific Discovery, and people might have spoken loftily about the value of science, but the value of gold and silver was more highly prized. When they stepped ashore, explorers and the scientists who accompanied them had their eyes on the ground, looking for the valuable and the unusual. Some also had their eyes on the skies, using astronomy to calculate the exact dimensions of the world. This was a valuable tool of knowledge for empires that were intent on measuring the world and carving it up.

There was another reason why empires aspired to pursue scientific objectives in the Antarctic. There was prestige to be gained in making hard-won scientific discoveries and announcing them to the world. European countries competed among themselves to be regarded as the fount of scientific knowledge. The United States was driven by a similar imperative. Its citizens had dreams of greatness for their expanding nation and regarded scientific discovery in the Antarctic as one way in which they might measure up against the old empires and proclaim their stature to the world. Commerce, science, and national prestige had become natural companions, and they would remain so in subsequent centuries.

Why was the burst of exploration activity in the 1840s followed by half a century of relative disinterest?

The voyages of Wilkes, Dumont d'Urville, and Ross, as well as the occasional forays by sealers, had filled in great swaths of the map. Thousands of miles of coastline south of Australia and New Zealand had been explored, and parts of the Antarctic

Peninsula and its nearby islands had been charted. However, these activities were mostly done in a rudimentary way by explorers in sailing ships whose captains and crews were sensibly cautious about venturing too close to a coast that was guarded by icebergs, shifting pack ice, strong currents, and hidden rocks.

Despite sporadic attempts over the seventy years since Cook had first sailed south, no explorer had managed to find a passage that would take their ships to the South Pole itself. By the early 1840s, explorers and sealers had only seen small stretches of coastline. Although Wilkes had called his sightings "Antarctica," he had no idea how big his continent might be. Moreover, Ross had shown that part of the "land" that Wilkes had claimed to have seen was just open sea. Nineteenth-century mapmakers still had no way of knowing how to fill in the most southerly parts of the world and were forced to use dotted or straight lines to connect the stretches of known coastline.

The continuing uncertainty about the South Pole might have been expected to spark a rush of polar explorers intent on determining whether there really was a continent or just a scattering of large islands and what the nature of any land mass might be. However, the Wilkes expedition had ended in recriminations between its officers and a court-martial for the hapless and high-handed Wilkes, while the failure of Ross to reach the South Magnetic Pole minimized the interest in Britain.

The impetus for further exploration was also muted by the continuing failure to find anything of sufficient commercial value to help justify the considerable cost of dispatching an expedition. Science was simply not sufficient justification by itself. The formidable ice cliff that stretched along the coastline seen by Wilkes, Dumont d'Urville, and Bellingshausen was a barrier to seals as well as to man. Not even the penguins could ascend its heights. As such, it was difficult to drum up financial support for a polar expedition that would only have science as its objective.

Polar explorers were also distracted during the latter half of the nineteenth century by the quest to find a Northwest Passage across the Arctic, which would link the North Atlantic and North Pacific Oceans and provide a sea route between Europe and the burgeoning markets of East Asia. The search had been begun by the English in the late fifteenth century and would take more than four centuries before meeting with limited success. When a British expedition led by Sir John Franklin disappeared in the Arctic in 1845, there were several attempts over subsequent decades to find some trace of his two ships. Although the remains of some of the crew were found in the 1850s, it was not until 2014 that one of the sunken ships was finally discovered near King William Island off the north coast of Canada, while the second ship was discovered nearby in 2016. The fruitless searches for Franklin's ships in the second half of the nineteenth century had focused public and government attention on the North Pole rather than the South.

There were also attempts to reach the North Pole in the 1890s. The most notable was mounted by the Norwegian explorer Fridtjof Nansen in 1893. He sailed his small vessel *Fram* into the Arctic Ocean with the intention of having it caught by the ice and taken by the current sufficiently close to the North Pole for him to reach it on foot with the assistance of dogs. The brave plan almost worked. The ice-bound ship drifted in the desired direction but did not get quite close enough for Nansen then to reach the North Pole on foot. It remained as a tantalizing quest that transfixed polar explorers for fifteen years, until two American expeditions claimed to have raised the American flag at the North Pole. Frederick Cook declared he'd done it in 1908 and Robert Peary said he'd achieved the goal in 1909. Although most observers at the time supported Peary, researchers have since discredited both their claims.

The exploitation of whales in Antarctic waters would have given a boost to exploration, but whalers stayed clear of Antarctica for most of the nineteenth century. There were still sufficient whales in the northern hemisphere and in the warmer parts of the southern hemisphere, where the dangers of the hunt were fewer than in the far south. Chasing and harpooning a whale amid the Antarctic ice pack, while perched on the prow of an open rowboat, would have been a particularly perilous job. It took the invention of explosive harpoons and steamships before whalers returned to the Antarctic, and a fresh bevy of explorers sailed in their wake.

2

THE RACE FOR ANTARCTICA IN THE TWENTIETH CENTURY

What caused a boom in Antarctic whaling in the late nineteenth century?

The 1860s and 1870s saw a decline in whaling when kerosene largely replaced whale oil for heating and lighting. The resulting drop in profits caused fewer whaling fleets to set out to sea. The American whaling fleet was also hard-hit by the Civil War, and by the 1880s there were hardly any American whaling ships left. British whalers were also much reduced in number. Their place was partly taken by the Norwegians, who had a long history of whaling and were well placed to benefit when serious money again started to be made from whaling in the last two decades of the nineteenth century.

There were several reasons for the reversals of fortune, which effectively removed the sanctuary that whales had enjoyed in Antarctic waters. The development of ships with ice-strengthened wooden hulls, which were driven by steam engines, helped tip the balance against the whales. Previously, sail-driven whaling ships had launched open rowboats when whales were sighted. With the sailors heaving on their oars, and with one man standing at the prow with his hand-thrown harpoon at the ready, the whales would be chased down and harpooned whenever they surfaced within reach. It was difficult, dangerous, and often unsuccessful work, particularly if a large whale dived to the depths with a harpoon lodged in its flank or if it simply sank when it died.

That all changed when fast and maneuverable whaling ships were able to chase down their quarries and shoot them with harpoon guns of increasing sophistication. The targets were not only the right whales but also the faster rorqual whales, such as humpbacks and fin whales. By the 1870s, whaling ships were being fitted with a harpoon gun on their bow that could fire a harpoon with an explosive head. The harpoon gun had been invented in the late 1860s by the Norwegian Svend Foyn, and its design gave the harpoonist a much greater success rate, particularly since steam-driven ships could outpace the faster whales. The ships had other important advances, since they used a steam-driven winch to drag the dead whales to the side of the ship, where a compressor injected air into the carcasses to keep them afloat.

The development of new killing methods coincided with the development of new uses for whale oil, which saw the price once again reach stratospheric heights. The most important new use was for explosives after dynamite was invented by the Norwegian Alfred Nobel in 1867. Glycerine from whale oil was a core component. Gelignite, cordite, and other nitroglycerin-based explosives soon followed. This caused a surge in the killing of whales in the North Atlantic, which led to a steep decline in whale numbers and forced whalers to look for new hunting grounds. By the 1890s, whalers began turning their attention to the Antarctic.

The wealth from whales increased even more in the early 1900s after a German chemist discovered a way of treating the strong-tasting whale oil to remove its taint. The successful hydrogenation of whale oil allowed it to be used for margarine and saw it quickly adopted as an ingredient for baking and other culinary purposes. Together with its use in explosives, whale oil once again became a highly prized product whose control was eagerly sought by the major nations of the world. As a result, the curtain of indifference that had hidden

Antarctica for half a century was swept away by the pursuers of profit and national glory.

How did the British and Norwegians come to dominate Antarctic whaling?

Britain sent a scientific expedition to the Antarctic in 1870 as part of an investigation of the world's ocean depths. With six scientists aboard, the steam-and-sail-powered HMS *Challenger* called at several sub-Antarctic islands before making a brief survey along the Antarctic coastline south of India. By dredging the ocean floor at regular points in the Southern Ocean, the scientists retrieved rocks that had dropped to the depths from melting icebergs. The nature of the rocks, which corresponded to rocks from other continents, suggested the existence of an Antarctic continent. But the rocks could not prove whether there was a single large continent or, as some suspected, two continents separated by an open channel. It was in this imaginary channel that whales were supposed to be in abundance.

Not to be outdone by the British, the Germans sent an expedition with the aim of looking for whales and seals and surveying some of the partly charted seas. In 1873, the German Polar Navigation Society dispatched the steam-driven whaling ship *Grönland* under the command of Eduard Dallmann. Heading for the Antarctic Peninsula, Dallmann hoped to find right whales and fur seals. While some useful survey work was done on the coastline and islands, Dallmann didn't encounter any right whales or find fur seals in commercial quantities. Consequently, there was no follow-up voyage by German whalers.

By the time whalers ventured south in the 1890s, the whalers of England and the United States had been surpassed by the Norwegians, who had maintained a strong whaling industry and whose ships were built for polar seas. The Scots also maintained a small whaling fleet out of Dundee, where whale oil was used mainly for softening jute for rope. With steam-driven ships now in ascendance, there was little appetite in America for the massive investment that was required to restart its whaling industry. The old whaling men and their financiers had left the industry, and there were few men left in the ports of New England with the expertise or desire to set off to distant seas to hunt for whales.

It was mainly therefore the enterprising Scots and Norwegians who mounted a serious investigation of Antarctic seas. In 1892, a Dundee whaling company sent four of its wooden sailing ships to the Weddell Sea, with auxiliary engines to help them work amid the ice-choked waters. The Norwegians responded with a smaller expedition of their own, with the Sandefjord-based whaler Christen Christensen dispatching his double-masted sailing boat *Jason* in pursuit. Under the command of Carl Larsen, the *Jason* was equipped to hunt for the slow-moving right whales. But neither the Scots nor the Norwegians found anything but the faster-moving rorquals, which they were ill equipped to kill. Neither did their ships return with commercial quantities of sealskins and oil.

Undeterred, both the Scots and the Norwegians tried again the following year. Again, they were unsuccessful. Another Norwegian expedition set sail in late 1893, hoping that they might find the lucrative right whales in the seldom-visited Ross Sea. Financed by the now elderly Svend Foyn, the expedition ship *Antarctic* searched fruitlessly for the whales that Ross had reported were there in vast numbers half a century earlier. There were plenty of other whales, but they were too fast to be caught. Despite this, the expedition was notable for landing seven of its men on the shore at Cape Adare in 1895, where thousands of bemused penguins watched the interlopers erect a Norwegian flag. Their feat showed that it was easy to go ashore and establish a base for exploring the unknown territory.

These successive expeditions did not fail because there were no whales to be found. They were present in great numbers, but they were the faster blue, fin, sei, and humpback whales that couldn't be caught with slow sailing vessels. Although coal-fired vessels were used to pursue whales close to the North Sea ports of Norway and Scotland, they were too small to carry sufficient coal to the Antarctic. Nor were there any whaling stations where the carcasses could be dragged ashore for boiling down. Unfortunately for the whales, that didn't mean that the whalers abandoned their hunt in the Antarctic. They just adjusted their methods to suit the challenges posed by the faster whales.

Why did the Antarctic begin to take on a strategic significance during the early part of the twentieth century?

In 1900, most of the Antarctic remained beyond the reach of whalers, whose hunting activity was limited by the need to tow freshly killed whales for processing at shore stations on the islands off the Antarctic Peninsula. They had to be flensed of their blubber as fast as possible to preserve the quality of the oil produced from boiling it down. The longer it took for processing to begin, the lower the quality of the oil and hence the price it could fetch. The introduction of factory ships changed all that. Instead of having to hunt within two or three days' sailing distance from a shore station, whalers were able to hunt over a much wider area. Working together, a factory ship could even range as far as the Ross Sea, so that its fleet of small, steam-driven whale chasers could hunt in those more distant and difficult waters. The dead animals would be brought to the factory ship for processing, with the factory ship also providing food and coal to replenish the whale chasers. As wireless sets were gradually adopted from about 1910, ships could keep in contact with each other and hunt with even greater freedom.

It wasn't just about whales. The intensifying imperial rivalry, which would eventually erupt in the Great War of 1914–1918, was also felt in Antarctica. Even though the exact outline of the continent remained unclear, Antarctica was assuming a strategic significance for the Great Powers of Europe. The use of whale oil for explosives and food ensured that the Antarctic whaling industry was of vital interest to the rival empires. So, too, was the control of the islands that lay athwart the Drake Passage that connected the Pacific and Atlantic Oceans. With their separate colonies in the southern hemisphere, and with the growing rivalry between their navies, both Germany and Britain were anxious to ensure access through the passage for their ships. Britain was better placed to do so. It had whaling stations on South Georgia, controlled the Falkland Islands, and had territorial claims to other sub-Antarctic islands.

Why did the increased whaling activity lead to renewed interest in Antarctic exploration?

On board one of the four Dundee whaling ships dispatched south in 1892 was a young doctor, William Bruce, who went as an amateur scientist and returned "ravenous"—as he put it— for further voyages to the Southern Continent. He wanted to do it for the sake of science and empire. If Britain wasn't willing, he proposed that Scotland should mount an expedition of its own. Norwegian expeditions also went hunting for whales in the 1890s and likewise returned with reports of the scientific booty that could be found there.

One of the Norwegian crew members was Carsten Borchgrevink, who rushed back to London so that he could attend the 1895 International Geographical Congress and argue for further expeditions to the Antarctic. The congress was presided over by a former British naval officer, Clement Markham, who didn't need much convincing about the need for more expeditions to the Antarctic. Under Markham's chairmanship, the congress called for them to be sent before the coming end of the century. Mounting purely scientific expeditions would bring prestige to nations and empires, while their geographical discoveries would assist whalers planning further speculative ventures in Antarctic seas.

In the event, the British and German geographers were beaten by the upstart Belgians, who sent a mostly private expedition to Antarctica in 1897. Led by the naval lieutenant Adrien de Gerlache, the expedition used a former Norwegian whaling ship, renamed *Belgica*, and a crew of mixed nationalities, including Roald Amundsen, who would eventually launch his own expedition, and the American doctor Frederick Cook, who would also later mount an expedition of his own to the North Pole and falsely claim to have reached it. Exploring the western coast of the Antarctic Peninsula, the *Belgica* was caught by the ice and held fast for thirteen months before a channel was finally cut through the ice to the sea. Their perilous experience might have dissuaded other explorers had the expedition not returned with a rich scientific haul and stories of having seen countless penguins, seals, and fin whales. These results whetted the appetites of whalers, scientists, and adventurers, who each had their own motives for wanting to head south in search of profit, science, or fame.

What other natural resources were regarded as being of potential value?

It wasn't just whales that were sought in the Antarctic. After being almost exterminated by the 1820s, seal numbers had recovered to such an extent that they were again being regarded as a resource ripe for exploitation. In his account of the *Belgica* expedition, Frederick Cook claimed to have seen large numbers of seals, penguins, and cormorants. Indeed, the penguins had provided much-needed food when the ship was stuck in the ice for more than a year. Cook believed that they could similarly provide food for workers in a fur industry on the sub-Antarctic islands, suggesting that Inuit people could be sent from the Arctic to the South Shetland Islands to manage a fur industry based on Arctic animals such as polar bears and foxes. His vision was never implemented.

There was also another possibility that excited visions of easy wealth. The discovery of gold in Alaska in the 1890s suggested that similar valuable metals or precious stones might be found in the Antarctic. The British explorer Ernest Shackleton certainly thought so and had that firmly in mind when he embarked on his expedition in 1907. Earlier thoughts of Antarctica being a useless wasteland were set aside in favor of frantic speculation about the wealth that the partly discovered Antarctic might contain.

What drove explorers to the Antarctic during the early part of the twentieth century?

As we've seen, the arrival of whalers toward the end of the nineteenth century reminded the world that there was still much to be discovered. So much was unknown that some atlases still showed Antarctica as two land masses with a large sea separating them. There was much to be done before the Antarctic would be known in all its sublime majesty. Attracted by the challenge, geographical societies, newspapers, and governments began turning their attention toward the South Pole. That might have led to sober, carefully planned programs of exploration based upon cooperation between different nations. However, the increasing imperial rivalry in the early twentieth century led instead to a frenzy of competing expeditions. Explorers and their supporters wanted to make their own separate, national marks on the continent, which would culminate in the race to be the first to the South Pole.

The English led the way. The reputation of their empire had been tarnished by the poor performance of the British army in the Boer War at the turn of the twentieth century, and they wanted to prove their prowess, courage, and masculinity to the world. Clement Markham, now the president of the Royal Geographical Society, was England's foremost advocate for an Antarctic expedition. As a naval officer, Markham had been aboard the ship that had scoured the Arctic in 1850, looking for

Franklin's lost expedition. By the 1890s, Markham had become an indefatigable campaigner for Britain to lead the world in uncovering new lands in the Antarctic. He said he was doing it to uphold his nation's naval leadership. As he explained in his memoirs, he wanted the exploration done by up-and-coming naval officers, who would be able to gain experience and test their courage and resourcefulness. However, he struggled to convince the British government to provide funding for the venture.

There were other ways of financing Antarctic exploration. The whalers and sealers had long been at the forefront of discovery, while the newspaper magnate Sir George Newnes had also shown how an expedition could boost the profits of a popular newspaper. In 1898, Newnes had thrown his support behind an expedition led by the Norwegian Carsten Borchgrevink, who sailed the ship *Southern Cross* to Cape Adare on the Ross Sea. He had gone ashore there several years before, and this time erected two huts, above which he flew the British flag. The men spent the winter on the ice, collecting whatever scientific specimens they could find. While Borchgrevink likened it to a "pioneer settlement," one of his gloomier companions described the Antarctic as being "enveloped in an atmosphere of universal death."[1]

When the *Southern Cross* returned in 1900 to collect him, Borchgrevink could hardly claim his expedition had been much of a success. Although he had shown that it was possible to winter on the ice, he wanted to announce a greater achievement. To do so, he sailed the *Southern Cross* to the southern extremity of the Ross Sea, where he and a party landed ashore with a sledge and a team of dogs. Making a brief push south to 78° 50' S, he declared that his expedition had outdone the 1840 expedition of James Clark Ross by venturing further south than anyone had done before. Unfortunately for Borchgrevink and Newnes, the outbreak of the Boer War in 1899 overshadowed the news from Antarctica. Nevertheless, the *Southern Cross* expedition showed how private philanthropy and newspapers could provide vital financing for future expeditions. Borchgrevink also pointed to the South Pole as the goal to which those expeditions should aim.

The Germans were not interested in racing to the South Pole from the Ross Sea, an area in which a succession of British expeditions had already made their mark. Far better, they thought, to head for a relatively untouched area where new scientific and geographical discoveries might be made. It would be a feather in the cap of the rising German empire if its expedition could show its scientific superiority and its men could display their prowess by exploring an entirely new area of the Antarctic, from where they might also be able to launch a bid for the South Pole.

More importantly, by concentrating their attention elsewhere, the German explorers might be able to answer a great question that still confounded geographers, as to whether or not the Antarctic was a single continent. The explorer Karl Fricker argued that Germany would thereby confirm to the world that it was the "Nation of Thinkers and Investigators."[2] The task was entrusted to the Arctic explorer and professor of geography at the University of Berlin Erich von Drygalski, who believed the exploration of Antarctica was a matter of national honor.

Under Drygalski's leadership, the German ship *Gauss* headed south in 1901 with a crew of scientists and naval officers, aiming for that unseen part of the Antarctic coastline that lay roughly south of India. It was there that Drygalski hoped that he might be able to discover whether there was a strait separating Antarctica into two land masses. He couldn't find one. What was worse, the *Gauss* was caught by the sea ice about 90 kilometers (40 miles) from the coast and remained stuck throughout the winter of 1902. Despite their predicament, the Germans made numerous observations and collected sufficient specimens to fill twenty scientific volumes. Still, they didn't return with any results to grab the attention of the world.

After the expeditions of Ross and then Borchgrevink, the British regarded the Ross Sea region as their own and planned a very different venture from other expeditions. Rather than a voyage of discovery, they would take up the challenge set by Borchgrevink: to spend a winter housed on the ice before setting out overland to reach the South Pole and/or the South Magnetic Pole. Their success would be measured in the size and number of their newspaper headlines rather than in the number of scientific volumes they could compile on their return.

With the mostly privately funded expedition being cast in the context of Anglo-German rivalry, a reluctant British government finally provided some financial support. The leadership was entrusted to a young naval officer, Robert Falcon Scott, who was lauded by Markham in his memoirs as a perfect gentleman. Scott's ship *Discovery* left London's docks in July 1901, with Ernest Shackleton aboard as one of the junior officers. It would be the beginning of a bitter rivalry between the two British explorers.

Why did the public become so captivated by the feats of the Antarctic explorers?

It was the start of a new century and the last of the world's undiscovered wonders, the location of the South Pole, was going to be revealed in a race between the Germans and the British. In the event, neither expedition got anywhere near the South Pole. While Drygalski's ship was caught by the ice far offshore, the (in fact) better-situated Scott and Shackleton expedition failed even to get off the Ross Ice Shelf with their dog-drawn sledges. By the time they ended their attempt, starvation and scurvy had nearly killed them.

Scott had gone further south than anyone else before, reaching just beyond 82° S, but he was still about 800 kilometers (500 miles) from the South Pole. Although he remained for another year and led a further sledge journey, Scott headed more than 550 kilometers (300 miles) southwest rather than trying again to go south toward the South Pole. This time, Shackleton didn't accompany him. Relations between the two men had broken down during the first sledge journey, and Scott sent Shackleton home on the relief ship. It was ostensibly for health reasons, but Shackleton regarded it as a slight and an embarrassment. Their relationship never recovered.

Although Scott got nowhere near the South Pole on his first expedition, the expedition had made him the preeminent English explorer of the Edwardian era. However, he would have to do much more if he was going to cement his reputation and achieve promotion to the rank of admiral, which he craved. Inevitably, that meant he would have to launch an assault on the South Pole before another expedition got there first. To Scott's chagrin, it was the upstart Shackleton who was the first to attempt it.

Having failed to create an alternative career for himself, Shackleton announced in 1907 that he would lead a new expedition that would attempt to reach both the South Pole and the South Magnetic Pole. Without support from Scott or Markham, the money Shackleton raised was barely enough to cover basic costs. He hoped that success would bring him both fame and fortune, telling his long-suffering wife that it would bring "ample money . . . for us to live our lives as we wish." Instead of relying upon dogs, as Scott had done, he would take horses as well as a car that he intended to use on the relatively smooth Ross Ice Shelf. Shackleton believed that horses would be able to pull a greater weight than dogs and provide more food when their job was done. However, as he would find out, they also required more food to keep them going, while the car broke down in the cold.

Shackleton paid little attention to science. Success would be measured instead on his impecunious expedition reaching the poles. That's what would delight the public and the British government alike. And he came close to doing so. Setting out from McMurdo Sound in late 1908, he succeeded in crossing the Ross Ice Shelf and finding a way up the treacherous Beardmore Glacier, where the last of his horses fell to its death down a crevasse. The remainder of the 1,200-kilometer (750-mile) journey across the polar plateau to the South Pole would have to be done by man-hauling the sledges.

By sheer force of will, Shackleton and his companions almost reached their goal. It was only their dwindling food supplies that finally forced Shackleton to concede on January 4, 1909, that the South Pole was beyond them. Pushing on in a last desperate dash to get as close as possible, Shackleton claimed to have gotten just 97 miles from his objective. Then they headed back. As difficult as the decision had been, their lives depended on it. As Shackleton later explained to his wife, it was better to be a live donkey than a dead lion.

His expedition's simultaneous attempt to reach the South Magnetic Pole also fell short. The three-man team, including the Australian geologist Douglas Mawson, had to trek their way along the Ross Sea coastline before striking westward to where the South Magnetic Pole was estimated to be. As with Shackleton's journey, it failed to reach its objective. Shortness of food caused the party to head back to the coast before they could determine the precise location of the Magnetic Pole. They took a photograph of themselves anyway, claiming to have reached it.

Although they had not reached the South Magnetic Pole, their 2,000-kilometer (1,200-mile) trek would remain for decades as the longest unassisted journey across the Antarctic ice. It also made a hero of the hitherto unknown Mawson, which gave him an opportunity to later mount an expedition of his own. As for Shackleton, he had blazed a path almost to the South Pole and would have to watch in frustration as other explorers followed in his footsteps.

With the British public captivated by the prospect of another assault on the South Pole, and wanting a British expedition to get there first, Scott could raise ample support for a renewed attempt. When some commentators criticized his new expedition for being a mere "pole-hunting exercise," which didn't merit the spending of public money, Scott responded by taking ten scientists with him. It was the greatest number of scientists that had ever gone on an expedition to the Antarctic. Rather than dogs, Scott would follow Shackleton's example and take horses for the first part of the journey before relying on the brawn of his men to haul the sledges up the Beardmore Glacier and across the gently rising expanse of the polar plateau. When Scott's ship, the *Terra Nova*, left London in May 1910, the assembled guests were told that the expedition would "prove once again that the manhood of the nation is not dead."[3] Such sentiments excited great public interest, which turned into frenzy when the Norwegian explorer Roald Amundsen declared that he was also heading for the South Pole and would also launch his attempt from the Ross Sea.

It wasn't just a two-expedition race. There were also German and Japanese expeditions setting their sights on the South Pole, making it also a race between the representatives of the aging British Empire and the upcoming empires of Germany and Japan. There was a further threat to Scott's ambitions in the person of Douglas Mawson, who declined an invitation to be a scientist on Scott's expedition and chose instead to lead an expedition that would have science, rather than "pole chasing," as its raison d'être.

It was the experienced polar explorer Amundsen whom Scott most feared. He didn't learn of the Norwegian challenge until just before he left New Zealand for Antarctica. It had come as a shock, since most of the world had believed that Amundsen had been aiming for the North Pole. Only when he was at sea in the mid-Atlantic did the Norwegian announce that he was heading south. Unlike Scott, he would rely on skis and more than a hundred dogs to haul his sledges before killing them for food on the return journey.

With the lightly equipped Amundsen hot on his heels, Scott pinned his hopes on three motorized sledges with caterpillar tracks to get him across the several hundred miles of the Ross Ice Shelf to the start of the Beardmore Glacier. If he could get onto the polar plateau first, Scott thought he would have the advantage. However, Amundsen landed on the edge of the Ross Ice Shelf nearly 100 kilometers (60 miles) closer to the South Pole than Scott. And he found another glacier to provide access onto the plateau.

Leaving nearly three weeks earlier than Scott, and with the help of the dogs, Amundsen and his four skiing companions made quick progress. Killing the dogs one by one and feeding their carcasses to the remaining ones, the Norwegians reached the South Pole by mid-December 1911. Leaving behind a tent, above which flew a Norwegian flag, Amundsen headed back triumphantly. The expedition had been so well organized that they arrived back fatter than when they had begun and with eleven dogs still able to pull their sledges.

From start to finish it was a race in which neither side ever caught sight of the other. But there was no mistaking the taunting sight of the Norwegian flag that the exhausted Scott and his four companions saw as they approached the South Pole a month after the Norwegians had left it. "Great God!," wrote Scott, "this is an awful place and terrible enough for us to have laboured to it without the reward of priority."[4] Scott still thought he might rescue something from the ruins of his expedition if he could get back first to Hobart in Tasmania and radio his achievement to the world, complete with photographs to prove it. Instead, he and his four companions died from starvation and the intense cold of the worsening weather on the return journey. The news of their terrible fate would take months to reach the world, while Amundsen returned to revel in the plaudits of the public, although they would be grudgingly given in Britain.

Why did Roald Amundsen succeed where others failed?

With Shackleton having shown the way, it shouldn't have been difficult for Scott to follow in his footsteps and get to the South

Pole first. However, Scott had made some fateful decisions that would weigh down his expedition and doom its members. One of the three motorized sledges was lost as it was being landed on the ice, and it plunged to the depths below. It wouldn't have made much difference. Operating the sledges, with their new-fangled internal combustion engines, proved too difficult in the extreme cold, where temperatures could descend to below -50°C. The two surviving machines were abandoned within 80 kilometers (50 miles) of the base camp.

The choice of horses rather than dogs also proved a fateful decision. It was prompted partly by an English dislike of having to kill dogs for food, but also because it was believed that the horses would perform better at hauling the heavy loads that were required to establish food stores at designated distances across the Ross Ice Shelf. To Scott's dismay, the horses struggled to make their way through the heavy snow, and as with Shackleton's earlier experience, they weren't able to climb their way up the crevasse-ridden glacier to the polar plateau. Although the horses could be killed and their flesh eaten, they couldn't be fed to each other, as dogs could, and much of the space on the sledges had to be taken up with their bulky food.

Scott had another reason for not taking dogs on the trek to the South Pole. Even though his expedition had taken some dogs, Scott left them behind and chose instead to man-haul the sledges up the glacier, which rose for more than 160 kilometers (90 miles) until it reached the relative flatness of the polar plateau about 10,000 feet above sea level. He believed that it was the manly thing to do once the horses had died or otherwise been dispensed with. Rather than skiing alongside dog-drawn sledges, as the lightly equipped Amundsen had done, Scott and his companions were forced to put on harnesses and drag the heavily laden sledges. Strapped into their leather harnesses, they were like pit ponies, with the terrible effort of hauling proving a fatal disadvantage. It required the men to have a much larger calorie intake than their dwindling food supply allowed.

By leaving for the South Pole much later than Amundsen, Scott loaded himself with an impossible burden. It meant that the Norwegians were almost back at their base camp when Scott had only just reached the pole. As a result of leaving later, Scott's return journey would see his party experiencing much colder weather conditions, including fierce blizzards that would keep them confined to their tent for days. Scott still felt impelled to divert for a day to collect 14 kilograms of rocks from a distant outcrop poking above the snow. The party not only lost time collecting the rocks but also weighed down their sledges with them. It may not have caused the tragic outcome, but it literally weighted the odds against success.

Scott needed something to flaunt when he faced the gaze of the world. Otherwise he would not only have been beaten by Amundsen but also have been outclassed scientifically by the other expeditions and regarded just as a failed "pole-chaser." He and his companions would not live to see the world's reaction. On March 29, 1912, they died in their tent on the Ross Ice Shelf, less than 20 kilometers (12 miles) from a food and fuel dump that might have saved them.

As a Norwegian and an experienced Arctic explorer, Amundsen was accustomed to the snow and cold. He and his men were good skiers, they knew how to control their dog teams, and they were not distracted by the demands of science. He had headed south with one aim, and that was to reach the South Pole first and be feted for it. Then he would be able to raise the capital for his real aim, which was to reach the North Pole. That ultimate objective animated his every move.

Although Amundsen beat Scott to the South Pole, why did Scott nevertheless emerge as the immortal Antarctic hero?

As the first person to reach the South Pole, Amundsen was rightly feted as a hero in his native Norway. He was the first person to reach the South Pole, and his small party had returned in fine physical condition, despite being at an apparent disadvantage in terms of equipment compared with Scott's resources. Yet the taciturn and reclusive Amundsen was not cut out to be a hero. He thought that his triumph would speak for itself. Time would show that he needed to be a showman like Shackleton or a writer like Scott if the world was going to comprehend the scale of his achievement.

It would also have helped had he been English. Many influential commentators, particularly in Britain, thought his behavior less than heroic. He was particularly criticized for keeping his intentions secret from Scott until he was well on his way to Antarctica. That was considered downright ungentlemanly. Amundsen's offense was compounded by launching his assault from the Ross Ice Shelf, which was regarded as belonging to Scott because of his prior exploration there.

Scott made a point of portraying himself as the very model of an English gentleman. He was a naval officer with ambitions to be an admiral. When all was lost and his death was certain, he saw to it that he would shape how the British public would regard his demise. His frostbitten companion Titus Oates had set a selfless example by stumbling out from their tent to die in the snow, so that his gangrenous feet would not impede the party's progress. Scott's account of that action set the scene for his own ending in the last days of March 1912. Rather than trudging on through the snow and collapsing in his harness along the track, Scott chose to stay in their tent. As the wind pushed relentlessly against the flapping canvas, and his companions lay about him dying in their sleeping bags, Scott calmly composed a series of letters to his wife, his friends, and the principal supporters of the expedition. Bereft of food and fuel, they lay in their sleeping bags for more than a week, contemplating their end.

The successful conquest of the South Pole was supposed to have brought financial security for Scott and his family. Although he had failed in his quest, he was determined they wouldn't suffer financially. In the confident expectation that their canvas-covered tomb and his final letters would eventually be found, he described how he had sacrificed his life for the nation and the greater good. Scott wrote his wife that she and their son "ought to be specially looked after by the country for which after all we have given our lives," while in another letter he described how they were "setting a good example to our countrymen" by facing their terrible predicament "like men." There is much truth in his depiction, though these last letters are also written self-consciously for posterity. It was calculated to resonate down through the ages, and so it has.

Scott's version of events dominated for more than half a century. Surviving members of his expedition wrote accounts that tended to embellish his growing mythic status, and historians tended not to deviate from the script. Scott had become a national idol, embodying the qualities of courage and endurance. His heroic status was invoked in the 1948 film *Scott of the Antarctic*, starring John Mills. Watching the film, war-wracked audiences of the shattered empire were reassured about the superiority of their race and the greatness of their nation's past.

Scott's place in the pantheon of British history seemed unassailable, and few were willing to rock the plinth on which his statue stood. That started to change in post-1960s Britain, when writers began to examine the planning and execution of the expedition. Gradually, a succession of critical studies began to chip away at the marble, with the most critical of all being a comparative study of Scott and Amundsen, published by Roland Huntford in 1979. The book was unflinching in its criticisms of Scott and its praise for Amundsen's leadership. Other writers followed Huntford's lead, often elevating Shackleton over Scott.

The case against Scott has recently taken a new turn, after several writers reexamined the evidence or brought their own Antarctic experiences to bear. The terrible weather conditions have been cited in his defense, with meteorologist Susan Solomon arguing in 2001 that Scott and his men would have returned alive if they hadn't encountered the exceptionally severe conditions. Of course, Scott wouldn't have been caught by those conditions had he employed Amundsen's methods. Then again, the tragedy might also have been averted had Scott's dog sledges played a more prominent part on the expedition, even if they had just come out to meet him on the Ross Ice Shelf, as Scott had instructed that they do. Although his defenders have grouped themselves around his remains, which were left in situ, it is unlikely that Scott will enjoy the unalloyed heroic status that he once did.

Who were the other Antarctic explorers of the heroic age and why have they been largely lost to history?

The first two decades of the twentieth century have come to be known as the "heroic age" of Antarctic exploration. In the popular imagination, it was a time when men pitted themselves against the elements without the benefit of modern technology. In fact, the age made use of all the cutting-edge technology that was available to them at the time. Ernest Shackleton took a car on one of his expeditions; Robert Scott took motorized sledges; Douglas Mawson took wireless sets and an aircraft, albeit damaged and without wings; and several explorers took tethered balloons to get a better view of the distant landscape.

Of all the expeditions, the race to the South Pole between Amundsen and Scott seemed to best encapsulate the heroic age. But there were several other expeditions during this period. Some were far more significant in a scientific and geographical sense than the race between Scott and Amundsen, while others were also defeated by the elements and their own deficiencies.

Perhaps the most scientifically minded explorer was Jean-Baptiste Charcot, a French doctor who led two expeditions. Charcot was already wealthy and didn't need to perform money-making feats. Neither was he motivated by national aggrandizement. Rather than exploring vast swaths of the continent, or racing to the South Pole, Charcot wanted to study a relatively small area in great depth. His first expedition in 1904 established a base on Wandel Island, off the coast of the Antarctic Peninsula, where the geography was relatively well known. Operating from a hut and on board his purpose-built ship, the *Français*, Charcot and his six scientists took regular meteorological and other observations and collected specimens from the land and sea. The findings didn't excite much public interest, but they were sufficient to fill eighteen volumes of scientific reports.

In a follow-up expedition in 1908–1909, Charcot headed back to the Antarctic Peninsula on a more lavish expedition in a new vessel that he cheekily named *Pourquoi-Pas* (Why Not). This time Charcot was partly funded by the French government. The gentleman scientist spent two summers charting 2,000 kilometers (1,200 miles) of coastline and collecting sufficient specimens and scientific data to fill twenty-eight volumes. Unlike other explorers, he didn't claim territory on behalf of his government, although he hoped that his work would prove that French civilization was the equal of any other.

Another scientifically minded explorer was the Scottish doctor William Bruce, who had gone to Antarctica on a Dundee whaling ship in 1892 and returned with the determination to mount an expedition of his own. He wanted to lead a British one but was unable to get sufficient support. Undaunted, he pressed ahead with a Scottish expedition, financed mostly by a Scottish industrialist and supported by the Royal Scottish Geographical Society. Science, rather than geographical discovery, would be at the core of the expedition, which Bruce hoped would bring honor to Scotland. Leaving in late 1902 aboard a refitted Norwegian whaler, renamed the *Scotia*, the Scottish National Antarctic Expedition headed for the Weddell Sea.

Like many explorers before him, Bruce found the ice of the Weddell Sea impenetrable and headed back to the South Orkneys, where he established a base on Laurie Island. Rather than a prefabricated wooden hut, Bruce spent six months erecting a solid stone building in the Scottish Highlands style, on top of which he flew the ancient Royal Standard of Scotland and the Cross of St. Andrew. It was all recorded using both still and motion-picture cameras. Not only was Bruce the first to take a cinematographic camera, he also took a phonograph to record the sounds of wildlife, including the plaintive cries of newborn seal pups before they were killed and preserved for science.

After refurbishing the *Scotia* in Buenos Aires and convincing the Argentine government to take over the base on Laurie Island as a permanent meteorological station, Bruce tried again in the southern summer of 1904–1905 to penetrate the ice of the Weddell Sea. This time he was more successful, reaching 74° S and coming in sight of the ice cliff that formed the shifting coastline of the continent. After charting 240 kilometers (140 miles) of the forbidding-looking cliff, Bruce headed home. On the pitching deck were several emperor penguins that he hoped would elicit the interest of both the public and scientists alike, but which died soon after their departure from their cold habitat. Of more enduring use were the meteorological and oceanographic data that Bruce brought back with him, while his base on Laurie Island would remain in Argentine hands and give that country a solid claim to the island's ownership.

Like the French and the Scots, a Swedish expedition might also have made its scientific mark had it not been so poorly funded and dogged by such awful luck that they were fortunate to escape with their lives. The expedition landed on Snow Hill Island in February 1902, just off the tip of the Antarctic Peninsula. It was led by Otto Nordenskjöld, who had earlier led an expedition to Tierra del Fuego and was the nephew of the famous Swedish explorer Adolf Nordenskjöld, who had sailed eastward across the Arctic Ocean from the Atlantic to the Pacific. The six-man expedition had no ship of its own and had to be dropped off by the Norwegian whaler Carl Larsen at the site of its future base. When Larsen returned to pick them up the following summer, ice prevented him from reaching the island. A rescue party sent on foot by Larsen was thwarted when it was unable to cross a stretch of open water to the island. It got even worse when Larsen's ship was crushed by the ice. The three separate parties were now stranded and had to survive the following winter before they were all fortuitously rescued by an Argentinian corvette. Despite it all, Nordenskjöld returned with sufficient scientific data to fill six volumes, along with more fossils to confirm the existence of an ancient continent with a temperate climate.

None of these expeditions received much attention. Neither did the expedition of Scott's German rival, Lieutenant Wilhelm Filchner, who was an experienced explorer of the Himalayan region. In a gentlemanly pact with the British, Filchner agreed to launch a bid for the South Pole from the Weddell Sea. While Scott grappled with his misfortune on the Ross Sea, Filchner's attempt to establish a base was thwarted when the ice shelf calved off into the sea, taking men, animals, and indeed his hut out to sea. Then his ship *Deutschland* was caught by the ice and forced to spend the winter of 1912 in its cold embrace, with its coal-fired boiler kept going with strips of oil-rich penguin blubber. These mishaps ended any German hopes of reaching the South Pole or ascertaining whether Antarctica was divided in two. Filchner's confinement on the *Deutschland* also meant that his scientific results would be largely oceanographic and meteorological and of no great public interest.

A little-known Japanese expedition also aimed to reach the South Pole. Under the leadership of an army officer named Lieutenant Nobu Shirase, it was designed to bring credit to the rising empire. In an appeal for support from his government, Shirase said that he intended to "expand the nation's territories" so that Japan could become "a rich and powerful nation."[5] On board his converted fishing boat, he took twenty-six dogs and two native Ainu men from Hokkaido, who were familiar with sledging and inured to the Antarctic cold. Shirase wanted to beat Scott and Amundsen to the South

Pole and planned to be based alongside his rivals on the Ross Sea. However, he arrived in mid-March 1911, which was just as the winter cold and darkness was about to hit. He retreated to Sydney to await the arrival of more men, money, and dogs. Setting out again the following summer, Shirase was a year behind his rivals and ill equipped to make little more than a token journey on the ice. However, his cinematographer recorded him raising the Japanese flag and claiming the region in the name of Japan's emperor. Shirase's ship had explored some of the Ross Sea coastline, and he had made a remarkably fast 275-kilometer (150-mile) dash across the ice with dog sledges and their Ainu drivers.

The greater challenge to Amundsen and Scott came from the Australian explorer Douglas Mawson. After having accompanied Shackleton in 1907, Mawson had intended to go south on another expedition mounted by Shackleton, this time as his chief scientist. When Shackleton postponed his expedition, he encouraged Mawson to mount one of his own and used his influence to ensure it was financed. Mawson wanted to distinguish his expedition from Scott's, declaring that it would be all about science. Also, unlike Scott, Mawson would focus his expedition on the discovery and exploration of previously unseen territory. He wanted to fill in the huge gap in the coastline south of Australia and to claim it for the British Empire. He also hoped to discover valuable minerals and to find a way of exploiting Antarctica's wildlife.

Mawson's original idea was to have three bases strung along several thousand miles of uncharted coastline. However, the ice cliff stretched along much of the coast and he was forced to have just two bases, the main one being on a rocky headland at Cape Denison. After erecting a hut in which they spent the winter of 1912, Mawson sent four parties in different directions. One went south in the vain hope of reaching the South Magnetic Pole. Another went west with a wingless aircraft acting as a makeshift, motorized sledge, only to break down and have to be abandoned. Another went east across the sea ice, charting the coast and collecting scientific specimens. The last party was led by Mawson, who took all the dogs to haul two sledges on an inland, easterly route. He and his two companions planned to travel about four hundred miles parallel to the uncharted coastline.

With the advantage of the dogs, Mawson should have been able to succeed. His companions were a champion Swiss skier and mountaineer named Xavier Mertz, and a young British army officer named Belgrave Ninnis. However, Mawson had not counted upon having to cross two massive glaciers. Ninnis was walking beside one of the heavily laden sledges when a hidden snow bridge collapsed beneath him, causing the sledge and its dogs to plummet into the depths. Mawson and Mertz were about three hundred miles from the safety of their base and had lost much of their fuel and food, along with their best dogs and tent. Rather than heading for the nearby coast, where there were seals and penguins to eat and where he might have encountered the coastal party or been rescued by the expedition ship, Mawson chose to head home by a route that would take them further inland.

It was a disastrous decision that was compounded by Mawson relying largely on the starving dogs for their food. With no fat on the dogs' flesh, the two men began suffering from protein poisoning, a condition that was little known at the time. Mertz ate more of the dog flesh than Mawson and soon sickened and died, leaving Mawson to complete the return journey alone. He barely made it back, only arriving after the rescue ship had left for home. Mawson was forced to spend another year in the hut, along with six expedition members who had volunteered to remain behind to search for him. Then he prolonged the eventual journey home by regularly stopping the ship to measure the changing depth of the ocean. Not believing in the theory of continental drift, he remained convinced that a land bridge must have once connected the continents of Antarctica and Australia and was determined to find it. But there was no land bridge to be found.

Somewhat to his surprise, and despite the deaths of Mertz and Ninnis, Mawson was treated like a hero in Australia and awarded a knighthood when he arrived in London. However, his book about the expedition failed to sell, and there were meager audiences for his lectures. The outbreak of war in Europe had diverted public attention from Antarctica. Yet Mawson's expedition had established an Australian claim to a third of the Antarctic continent and made him that nation's preeminent Antarctic hero. The scientific reports and observations took decades to publish, but they helped to buttress the Australian claim.

The outbreak of war not only overshadowed Mawson's achievement but nearly scuppered a new expedition by Shackleton, whose ship was just about to leave for the Antarctic when the war erupted in August 1914. Called the Imperial Transantarctic Expedition, Shackleton proposed to cross the continent, which he trumpeted as "the greatest Polar journey ever attempted." It was designed to outdo both Scott and Amundsen. Shackleton planned to land on the coastline of the Weddell Sea and cross the continent to the Ross Sea. To reduce the risk of failure, he sent another party to establish a base on the Ross Sea, from where they were instructed to set up a line of food and fuel dumps along the 2,400-kilometer (1,500-mile) route that he was planning to take.

It was all wasted effort. When Shackleton reached the Weddell Sea in January 1915, he was confronted by an ice pack that his ship, the *Endurance,* could barely penetrate. His attempts to do so ended in the vessel being caught fast by the ice. Like others before him, Shackleton thought he could wait out the winter. Instead, the grip of the ice became steadily tighter with each passing month until the *Endurance* was finally crushed and sank. Forced to take refuge on the ice, Shackleton led his men on a perilous journey by foot, with three small boats to traverse the ice pack and open sea, to the relative safety of Elephant Island off the Antarctic Peninsula. Seals and penguins provided food as they waited vainly for a passing whaling ship to sight them. No ship came, so Shackleton and several of his men took one of the small boats on a perilous voyage over some 1,300 kilometers (800 miles) to South Georgia Island. Landing on the island's southern coast, he crossed its snow-covered mountains to a whaling station on the island's northern shore. It was an extraordinary feat that was captured by the spectacular photographs of Frank Hurley.

Although all the men on the *Endurance* survived the expedition, the eight men of the Ross Sea party were not so fortunate, with one dying of scurvy and two others perishing in a blizzard. Like Scott, Shackleton had achieved nothing of any consequence, and his inspirational story of survival against the odds was overshadowed by the war. Only much later would his expedition come to be seen as the epitome of courage, fortitude, and leadership.

Had he realized that his reputation was secure, Shackleton might have rested content. But it wasn't just about reputation. The fortune he'd been pursuing still eluded him, and another expedition seemed the only way that he could achieve it. This time, he planned to take an aircraft to help with the exploration of some 3,000 kilometers (1,800 miles) of uncharted coastline in West Antarctica. Leaving in late 1921, Shackleton was dogged by difficulties with his ship, which forced him to lay up in Rio de Janeiro for repairs and delayed his arrival at South Georgia. The stress of it all caused the aging explorer to suffer an initial heart attack in Rio and a second fatal one on board his ship at South Georgia, where his body was buried in a grave overlooking the Norwegian whaling station. If there was such a thing as the "heroic era," Shackleton's death marked its end.

How did the use of aircraft change the equation?

The expedition of Douglas Mawson in 1911–1914 could have marked a historic shift in Antarctic exploration through his use of an aircraft. However, the loss of the plane's wings during a demonstration flight in Adelaide meant that it was relegated to being an improvised, motorized sledge. With his plan to use a seaplane, Shackleton's final expedition in 1922 might have reignited public interest in the Antarctic. He thought the aircraft could assist with the exploration and enable them to take spectacular aerial photographs of the hitherto unseen landscape. However, the expedition ended in disarray after Shackleton's death and the plane was never deployed, although one of the members of that expedition was the first person to fly an aircraft in Antarctica.

The Australian-born Hubert Wilkins had made a name for himself as an aviator and motion picture cameraman in the Balkan War of 1912–1913 and on the Western Front during World War I. He had been a cameraman on an expedition to the Canadian Arctic in 1913–1915 and also competed unsuccessfully in the 1919 air race between Britain and Australia. Wilkins was also a member of the little-known British Imperial Antarctic Expedition, which was meant to involve 120 men, a fleet of ships, and twelve aircraft, with plans to establish permanent British bases on the continent. Due to lack of funds and poor leadership by John Lachlan Cope, the expedition ended up being composed of just four men who were landed ashore by a whaling boat. Cope and Wilkins left their hut on the Antarctic Peninsula after one summer, leaving their two young companions to stay on for the winter of 1921.

Wilkins had then joined Shackleton's equally ill-fated expedition that same year, when again he had his hopes dashed of flying in the Antarctic. Nevertheless, he returned to London with a grand vision that would preoccupy him for decades. Wilkins proposed that a ring of meteorological stations should be established around the Arctic and the Southern Ocean to forecast the weather and even the climate of nearby continents. It was a scheme that was too grandiose for the times. Meanwhile, Wilkins shifted his attention to the Arctic, where he embarked on an ambitious plan to fly in search of a land that was meant to be located between the North Pole and Alaska.

Also heading for the North Pole was the American aviator Richard Byrd in an aircraft and the irrepressible Roald Amundsen in an Italian airship. They all failed, although Byrd would claim that his plane had reached the North Pole during his flight in 1926. Amundsen's airship crossed the Arctic from Europe to Alaska, but there was so much cloud cover that he was unable to sight the supposed land. As for Wilkins, he succeeded in crossing the Arctic from Alaska to Spitsbergen in 1928 and also failed to find any land. Despite persistent doubts about his achievement, Byrd was greeted with ticker tape on his return to New York and lauded by the U.S. Congress and American scientific societies. Further fame came in 1927 when he successfully flew across the Atlantic.

The Antarctic still waited to be conquered by aviators, however, who were by now confident that their aircraft could operate in intense cold and safely land and take off from the continent's icy surface. The stage was set for another race, this time by aircraft to the South Pole, with several aviators announcing their intention to take part. Along with Wilkins and Byrd, there was the wealthy American adventurer Lincoln Ellsworth who had been on the airship with Amundsen. There was also the Argentinian engineer Antonio Pauly and the British polar explorer Douglas Jeffery. In the event, Ellsworth, Pauly, and Jeffery were unable to get their expeditions off the ground, which just left Wilkins and Byrd. Aircraft manufacturers, scientific societies, and newspapers were keen to finance them, and the British and American governments wanted to lend their support in the hope that the respective expeditions would buttress their own territorial claims to part or all of Antarctica.

The media trumpeted the expeditions as a race between Wilkins and Byrd, each of whom was backed by a rival American newspaper company, yet the two expeditions were heading for different parts of Antarctica. Wilkins wanted to fly from the Antarctic Peninsula to the Ross Sea in the hope of finding suitable places to establish some of his permanent meteorological stations. His flight might also confirm whether Antarctica was one continent or two. Only when the flight to the Ross Sea was completed would Wilkins think about possibly going on to the South Pole, whereas Byrd just wanted to fly to the South Pole and be the first to fly over both poles. While Wilkins's flight would reinforce Britain's territorial claim to that large swath of the continent, Byrd intended to plant the Stars and Stripes at the South Pole and thereby lay the basis for an American claim to all the territory he could see from his cockpit and perhaps even the entire continent.

Taking off from an improvised dirt strip on Deception Island on December 20, 1928, Wilkins became the first person to see Antarctica from the air. However, his hopes of flying all the way to the Ross Sea were dashed by the shortness of his runway, which restricted the amount of fuel he could carry, while the need for his aircraft to use wheels rather than skis prevented him from landing anywhere else but back on Deception Island. Nevertheless, he managed to traverse much of the Antarctic Peninsula in a twenty-hour flight that took him over a strait that he mistakenly believed separated the peninsula from the rest of Antarctica. It seemed that there were two land masses after all. Wilkins took the news back with him to the United States, along with spectacular aerial photographs of the snow-covered peaks and plateaus. He informed the British that he had dropped a Union Jack at the furthest extent of his flight to reinforce Britain's claim to the peninsula.

In contrast to Wilkins's expedition, Byrd's was the most expensive ever mounted in Antarctica. With two ships and three aircraft, his party of eighty men were just establishing their base on the Ross seacoast when Byrd received news that Wilkins had completed his historic flight. Byrd was left to explore more of the continent and possibly be the first to reach the South Pole by aircraft. Rather than heading straight for the South Pole, in January 1929 Byrd began exploring the land to the east of the Ross Sea. It was territory that Britain aspired to own but hadn't yet done anything to claim. So Byrd claimed it for the United States and named the 20,000 square miles "Marie Byrd Land" after his wife.

With backing from the *New York Times* and *National Geographic*, and with a journalist along to report all his activities, Byrd garnered far more coverage than Wilkins. He had even sold the book and film rights. Given that Byrd had three aircraft and time was no longer critical, he covered more of the continent than Wilkins, who had returned in late 1929 for a second summer of aerial exploration. This time, Wilkins hoped to go much further, perhaps even reaching the South Pole. Instead, he was dogged by ill luck. Although he made more flights, they were shorter due to poor weather and failed to get anywhere near the Ross Sea, let alone the South Pole.

Byrd was ensconced in his Ross Sea base, named "Little America," when he received the news of Wilkins's return. Taking off for the South Pole on November 28, 1929, Byrd had a cameraman on board to film the flight for cinema audiences as well as a survey camera to film to the left and right of the aircraft's route of 1,250 kilometers (775 miles) to the South Pole. Doubts as to whether Byrd had reached the North Pole persisted, so he wanted to ensure that the world would actually witness him reaching the South Pole. The cameras were also designed to help establish an American claim to all the territory he could see from his cockpit, which encompassed far more than Scott, Shackleton, or Amundsen had managed to see at ground level, although an airborne claim would never be recognized by other nations.

Ditching some of his heavier cargo as the aircraft threaded its way between mountain peaks, Byrd struggled to lift the plane safely above the 3,400-meter-high (11,000 feet) polar plateau on which the South Pole is situated. Once he was satisfied that his quest lay below him, Byrd made no attempt to land. He simply dropped a Stars and Stripes through an opening in the plane's floor. The British and Norwegian flags that Byrd had taken to honor the achievements of Scott and Amundsen returned with him on the eight-hour flight back to Little

America, while the news of his achievement was radioed to the world.

Byrd had not only reached the South Pole but also flown over more than 400,000 square kilometers (155,000 square miles) of territory, much of it hitherto unseen. And he went on to explore much more territory from the air, concentrating on the previously unseen land to the east of the Ross Sea, all the while photographing it with the previously uncharted coastline in the background so that his maps would be more accurate and his territorial claim more certain. Byrd also sent out a sledging party to cross the territory on foot and erect a cairn of stones on a mountain in Marie Byrd Land, thereby confirming it as American territory.

The advent of aircraft in Antarctica inspired Mawson to return in 1929 to reinforce the British claim to the wide wedge of Antarctica located to the south of Australia. He wouldn't attempt to replicate his ill-fated dog sledging journey of 1912–1913. Instead, with a ship and a small seaplane, Mawson traversed several thousand miles of coastline, landing ashore wherever he could to raise the flag and sometimes to erect a cairn of stones. When the sea was sufficiently calm for his aircraft to take off, he also went aloft to drop flags on the continent in the hope that it would cement his claim to the Australian Antarctic Territory.

The Norwegians knew as much, if not more, about Antarctica than the explorers from any other nation. Their whalers had been hunting in the Antarctic for decades and had established permanent whaling stations, complete with boiling-down works and even a chapel, on its offshore islands. One of the most enterprising of the whalers was Lars Christensen, who used aircraft to give his fleet the edge over rival fleets in locating pods of whales. Linked to the fast whale catchers by radio, aircraft could sight and follow their quarry until they were within harpooning distance of the pursuing whale catchers. But Christensen didn't use aircraft just to catch whales. The Norwegians were upset by the British demand that they pay for the right to hunt in waters that the British claimed as their own. Christensen decided that the only way to defend his commercial interests was to make a territorial claim on behalf of Norway. Starting in 1927, and with the support of the Norwegian government, he used his whaling profits to do just that, intending to claim a massive wedge of Antarctica, stretching from 60° E to 20° W, comprising about a quarter of the continent.

In late 1927, Christensen sent the wooden sealing ship *Norvegia* to explore the mostly unknown coast and look for islands that could be useful to his whaling enterprise. He sent it back again in 1928, the same year that Byrd, Wilkins, and Mawson were making their own aerial forays, and in 1930 he also sent the two polar explorers Hjalmar Riiser-Larsen and Finn Lützow-Helm with a seaplane lent by the Norwegian navy. Like their rivals, the Norwegians used the aircraft to land at suitable places where a symbolic claim could be made by raising the Norwegian flag, photographing the ceremony, and leaving behind a cairn of stones. When they couldn't land, Norwegian flags were dropped from the aircraft onto the ice. It was a dramatic transformation from the time of Scott and Amundsen.

When did people know beyond doubt that Antarctica was a single continent rather than two or more large islands?

In the early twentieth century, some European atlases were still being published showing Antarctica to be composed of two continents with a vast sea between them. Other maps showed Antarctica as a single continent, with notional coastlines linking the various explored parts into an imagined whole. The use of aircraft might have settled the matter. However, it was impossible to be certain when looking down on a snow-covered surface whether it was covering solid land or sea ice. That was the mistake made in 1928 by Wilkins. After flying along part of its length, Wilkins thought he saw a frozen strait that separated the peninsula from the bulk of the continent. Millions of maps were produced in the wake of Wilkins's expedition and distributed to American children, revealing his discovery of a supposed archipelago that was actually the Antarctic Peninsula.

Another Australian explorer, John Rymill, helped to disprove Wilkins's sighting of a frozen strait. Rymill was leader of the British Graham Land Expedition, which did a detailed exploration of the Antarctic Peninsula in 1934. With curious penguins looking on, Rymill nudged his small ship *Penola* through the melting ice floes as he charted the peninsula's western coastline and painstakingly produced the most accurate charts to date. In doing so, he searched in vain for the frozen strait that Wilkins was reported to have seen. He couldn't declare conclusively that a frozen strait did *not* exist somewhere beneath the vast whiteness of the continent that connected the Atlantic and Pacific Oceans. Neither could Byrd make such a declaration after he returned to the Antarctic in 1934 and also looked fruitlessly for the supposed strait. Even a more systematic series of photographic survey flights by the U.S. military after World War II failed to find the strait. This reinforced the belief that Antarctica really was a single land mass, which was further reinforced by the first satellite photographs of Antarctica in the 1960s. These satellites would become more sophisticated and allow their instruments to probe beneath the surface of the ice to produce the first definitive maps of the entire coastline, thereby finally and irrefutably solving the centuries-old puzzle.

3

IMPERIAL RIVALRY

How did nations justify their claim to territories they were not occupying?

As the windiest, coldest, and driest continent on the planet, Antarctica poses great challenges for nations that aspire to own part or all of it. Most of the continent is covered by an immensely thick layer of ice that is sliding slowly but inexorably toward the surrounding seas. Apart from these natural challenges, there is no readily accessible source of fuel other than oil derived from boiling down seals or penguins. Most explorers relied upon coal and kerosene or other liquid fuel, which was costly and laborious to transport over such long distances. They couldn't live without these fuel sources since they were vital for heat, food, light, engines, and producing water from ice. It was just one of the challenges that seemed to make permanent settlements impossible. Yet international tribunals had allowed nations to claim territories that were too difficult to inhabit. Antarctica was regarded by many international lawyers during the first half of the twentieth century as being similar to such territories.

Based on this understanding, and with the world's strongest navy and a firm hold on the Falkland Islands, Britain simply declared in 1908 its ownership of the South Shetlands, the South Orkneys, South Georgia, and the Antarctic Peninsula. The declaration ignored the fact that Argentina had a permanent meteorological station in the South Orkneys, that Spaniards or Americans were discoverers of some of the islands, and that there was no official British presence on any of the islands or on the peninsula. Lawyers in the British Foreign Office deemed it sufficient for the British warship in the Falklands to occasionally visit the islands and raise the flag.

In justifying its ownership of unoccupied places, Britain pointed to its exercise of authority over them, mainly by levying a license fee on foreign whalers. Britain justified its fees as serving to protect the whales so they would not be hunted to near extinction as fur seals had been. By paying these fees, the Norwegians implicitly recognized British sovereignty over the extensive Falkland Islands Dependencies. Or so the lawyers in London assured the British government. That legal opinion persisted until World War II, by which time it had been proved possible to have permanent settlements in Antarctica.

Britain followed up its claim on the Falkland Islands Dependencies by declaring in 1923 its ownership of the Ross Dependency, which New Zealand then administered on its behalf. It was the region that Robert Scott and Ernest Shackleton had explored, and both explorers had acted as New Zealand postmasters and taken New Zealand stamps with them. Situated between 160° E and 150° W, most of the Ross Sea is ocean, with the land component comprising just 450,000 square kilometers (out of Antarctica's 14 million square kilometers). Although a handful of New Zealanders had been members of the Scott and Shackleton expeditions, no New Zealanders stepped ashore on the coast of the Ross Sea during the three decades after the declaration. Yet New Zealand was still able to collect fees from the whalers who hunted in those waters. Although the whalers of some nations, such as Norway, were willing to recognize the Dependency, the United States resolutely refused to do so for fear of jeopardizing its own potential claims to the region. New Zealand's sense of ownership was buttressed by its relative proximity to the region it was administering.

What were the effects of Richard Byrd's flights?

The first aircraft flights in Antarctica during the 1920s marked a momentous change in the exploration of the continent. Aviators could see in a single flight what could take months to see by dog sledge. And the United States was at the forefront of that exploration, with the arrival of Richard Byrd marking the beginning of renewed American interest. American newspapers had made money out of the race to the North Pole in the early 1900s and Byrd's later attempt to fly over the North Pole in the mid-1920s. Now the media money followed Byrd and his rivals as they sought to be first to reach the South Pole by air. The wonder and grandeur of Antarctica could be mined for newspaper readers, radio listeners, and cinema audiences. The excitement of an aerial race to the South Pole was portrayed in full-page aerial photographs and films of the ice-covered splendor.

Although Byrd promised that he would not encroach on British or Norwegian territorial claims, his behavior during the 1928 expedition started a process he couldn't stop. When Byrd dropped an American flag onto the South Pole, and then broadcast his feat to radio audiences across the United States from "Little America," it's not surprising that his action created a sense of ownership. Byrd would go on to describe Antarctica as "the world's last frontier," a description that had a special resonance for American listeners, and he gave American names to many of the continent's geographical features. His expeditions in the early 1930s led to strident calls by the U.S. Congress for an official territorial claim to be made over all those parts of Antarctica that could be said to have been discovered by American explorers—from the early sealers of the 1800s to the official American expedition of the 1840s and aviators of the 1920s and 1930s. The call found sympathy with President Franklin Roosevelt, who ensured that Byrd was provided with practical government assistance of various kinds, including permission to establish a branch of the U.S. Post Office in 1934 at Little America, where Byrd spent fourteen months.

Roosevelt also had strategic objectives in mind. Indeed, they were probably paramount in his thinking. He was conscious that Britain wanted to claim all of Antarctica for its empire and was determined to preempt the United States. Roosevelt also believed that American control of Antarctica, and in particular the Drake Passage, could provide the United States with a foothold from where it could launch its air power across the oceans of the southern hemisphere. In particular, it could provide an alternative reinforcement route to the American-controlled Philippines by way of South America, Antarctica, and Australia. Such a route could be essential in the event that hostilities with Japan cut the island-hopping path across the Pacific. When Japanese and German whaling fleets began appearing in Antarctic waters in ever greater numbers in the late 1930s, America's strategic interest in the continent became more compelling than ever.

What was Britain's claim?

By 1919, the value to be had from whaling and the strategic importance of controlling the Drake Passage convinced the British government to claim the entire continent for its empire. A British fleet had fought a naval battle with eight ships of the German navy near the Falklands in December 1914, sinking six of the ships and ensuring British control of the Pacific and South Atlantic. At the same time, the value of whale oil, which as noted was used for both foodstuffs and explosives, made the Antarctic waters a valuable asset. The British ambition to claim the entire continent wasn't only for economic and strategic reasons. The continent had come to exert a strong pull on the British imagination ever since Scott's tragic expedition to the South Pole, which helped create a sense of British owner-ship of the Ross Sea region where Scott and his companions had met their end.

Although Britain emerged from World War I scarred and exhausted, its seizure of German colonies saw the British Empire expand to its greatest-ever extent. Britain now had colonies, trusteeships, or self-governing dominions in every ocean and on every continent. How grand it would be if the British Empire could also include the entire continent of Antarctica, which was potentially capable of reinforcing Britain's naval supremacy in the oceans of the southern hem-isphere. Seals and penguins might produce some profit, and valuable mineral deposits could lie beneath the ice, though no one knew whether it would ever be economical to exploit them. Ordinarily, those uncertainties would have deterred Britain from claiming ownership of such a large land mass, but the small cost of establishing a territorial claim made it a risk worth taking. The continent didn't have to be garrisoned or even require settlements, just a declaration of British sover-eignty and the establishment of a British administration that could be exercised remotely. To achieve these goals, Britain simply needed to dispatch an occasional expedition and regu-larly raise the British flag at points around the coastline.

A junior minister in Britain's Colonial Office, Leo Amery, got things going. Amery had been a member of the British delegation at the Versailles peace negotiations in 1919, when Germany had been forced to relinquish any claim it had to parts of Antarctica based upon the discoveries of its expeditions. As the British Empire moved to absorb many of Germany's overseas possessions, Amery argued for Britain also to absorb Antarctica. He dismissed out of hand the territorial claims of any other nation, including France, Norway, and the United States, and convinced his colleagues to agree in principle to assert a British claim over the entire continent. In the first in-stance, Britain added the Ross Sea sector to their existing claim to the Antarctic Peninsula.

Cooler heads than Amery's prevailed. Foreign Office officials pointed out that they couldn't ignore the French claim to Adélie Land and that Britain could only claim those parts of Antarctica that its explorers had discovered. That would leave about half of the continent unclaimed until its unseen coastline could be explored and parties were sent ashore to raise the flag. It wasn't what Amery had wanted, but his promotion to the position of colonial secretary in 1924 allowed him to press ahead with his vision of creating a British Antarctica. He would do it by stealth, claiming the continent piece by piece until the only non-British sector was the French claim over Adélie Land. At least, that was his plan.

When dominion delegates—from Canada, Australia, New Zealand, South Africa, and elsewhere—assembled in London for the 1926 Imperial Conference, Amery told them of his secret scheme to gradually assert British sovereignty over the continent in the hope that "foreign Powers will acquiesce and that practically complete British domination may in time be established." Amery still wondered privately whether the continent itself would ever be of much value other than perhaps for, as he put it, "all round winter sports." It was just as well, because his hopes of annexing the remainder of the continent were quickly dashed.

How did other nations react to the British claim?

Britain may have forced Germany in 1919 to cede its right to claim territory in Antarctica, but other nations with Antarctic interests were not so accommodating. The French reminded Britain that the expedition of Jules Dumont d'Urville in the 1840s had discovered and claimed Adélie Land. France was determined to retain control of Adélie Land, though more than eighty years had passed since any of its citizens had been there. Britain was not prepared to rupture relations with its neighbor and wartime ally over such a minor issue and quickly agreed that Adélie Land rightfully belonged to the French. Britain secretly hoped that France might one day agree to swap it for a suitable island in the Southern Ocean.

Of all the nations with interests in the Antarctic, none had a greater contemporary claim to the continent than the Norwegians. Roald Amundsen had after all been the first to raise a national flag at the South Pole, and Norwegian whalers dominated the industry in the Southern Ocean. Yet the Norwegians paid a fee for hunting or stationing their factory ships in waters claimed as British, and they operated whaling stations on South Georgia under the authority of the British officials who administered the island as part of the Falkland Islands Dependencies. While the Norwegians submitted to British sovereignty, the whalers did whatever they could to avoid paying the fees. The development of huge factory ships with rear-loading ramps, which allowed whales to be easily hauled aboard and flensed of their blubber on open seas, meant that the Norwegian whalers could do much more of their hunting beyond the waters claimed by Britain and refuse to pay any fee at all. To buttress that newfound advantage, the Norwegians decided to establish their own claims to parts of the continent that were still unclaimed in East Antarctica.

Norway mimicked British methods by raising the Norwegian flag on any territory it wished to claim and then publishing maps on which Norwegian names would be featured. Wherever possible, huts with emergency supplies for shipwrecked Norwegian whalers were erected on exposed rocky outcrops. The Norwegians were fortified by their conviction that whaling was their "ancient inheritance" and that the polar regions were their natural home. After all, Norway stretched all the way into the Arctic, and the Norwegians had established settlements in ice-cloaked Greenland about a thousand years ago. With the full support of the Norwegian government, the whaler Lars Christensen made a determined effort from the late 1920s onward to claim the huge wedge of East Antarctica that lay to the south of Africa and India between 60° E and 20° W. Using a specially equipped expedition ship, as well as his whaling vessels, Christensen was intent on making a quarter of the continent Norwegian.

While the Norwegian gambit was an annoying distraction for the British government, the United States posed a far greater threat to Britain's ambition. With its control of the Falklands, South Georgia, South Africa, New Zealand, and Australia, the British government could make life difficult for a Norwegian whaling fleet operating in the Southern Ocean. That wasn't possible with the United States, which had historic claims of discovery in Antarctica and a commitment to the unilaterally imposed Monroe Doctrine of 1823, which aimed to exclude European powers from the Americas.

The United States contested the British claim to the Antarctic Peninsula, arguing that Nathaniel Palmer and other sealers of the 1820s had been the first to discover the peninsula, which the Americans named Palmer Land. The maps by William Smith and Edward Bransfield, which were put forward by the British to argue the primacy of their discovery, were rejected out of hand by Washington as forgeries or mistaken. The Americans also pointed to the discoveries and maps of the official expedition by Lieutenant Charles Wilkes of the 1840s, which had sighted a large stretch of the Antarctic coastline to the south of Australia. Although Wilkes mistakenly declared that a cloud bank was part of the Antarctica coastline, he was the first to suggest that he had discovered a continent and that it should be called Antarctica. As such, argued Washington, he should be credited as the discoverer of the continent. Such arguments consumed the attention of several British and American geographers in the 1920s and 1930s in the pages of scholarly journals and occasionally in the popular press. Extra fuel was added to the arguments with each discovery of a new document in the British or American archives, or as logbooks of long-departed whaling captains were unearthed in their attics.

While British and American geographers contested the veracity of the early maps, Washington encouraged Byrd to make incontrovertible claims to the territory east of the Ross Sea, which no one else had explored, and named it "Marie Byrd Land." During the 1920s and 1930s, Byrd was able to travel much further and faster than earlier explorers, and he was able to photograph his discoveries from the air rather than having to painstakingly survey them from the ground. The only problem was that aerial photographs did not provide a sufficiently accurate basis for the maps that were needed to support his territorial claims. He had to follow up his flights with sledging journeys across Marie Byrd Land, which allowed the height and location of mountains to be determined with greater accuracy. Yet he was still unable to produce satisfactory maps.

Apart from the maps, the establishment of a radio station and a post office at Little America indicated to the world that the United States was exercising administrative authority over the Ross Sea. Despite pressure from Britain and New Zealand, Byrd politely declined to ask New Zealand for permission to base his expedition there. His actions showed the world that America was creating the basis for later making an official claim to those parts of Antarctica that its explorers had discovered. This approach certainly alarmed the British government, since it drove a wedge between British claims to the Ross Sea region and Antarctic Peninsula and preempted Britain's ambition to own most of the continent.

Americans and Norwegians were not the only nations intent on claiming parts of Antarctica for themselves. Argentina and Chile argued that they had the strongest claims of any nation, particularly to the nearby Antarctic Peninsula. In 1904, Argentina had taken over the base of the Scottish National Antarctic Expedition on Laurie Island in the South Orkneys and ran a permanent meteorological station, with a small staff of meteorologists and radio-telegraphists remaining there even after the South Orkneys were made part of the British-controlled Falkland Islands Dependencies in 1908. The creation of the Dependencies was meant to ward off Chile and Argentina, both of which regarded the islands not as colonies but as an integral part of their respective national territories.

The fact that their territories overlapped meant that the South Americans could not present a united front against Britain. Until World War II, London was able to use the implicit threat of its naval might to hold off these challenges.

What was the so-called sector principle and how did it affect territorial claims?

On other continents, a river or a mountain range might be the agreed demarcation line between different nations. It was more difficult in Antarctica, where there were no long rivers and where other geographical features were mostly buried beneath miles of ice. Since it seemed impossible to establish settlements in the continent's interior, it was far simpler for nations to adopt the so-called sector principle, which had been accepted by some countries, although not the United States, for dividing up the Arctic. The sector principle used lines of longitude to delineate the areas claimed by different nations, with the areas encompassed by lines of longitude stretching all the way to the North Pole. When applied to Antarctica, it meant that the explorers from countries like Britain that accepted the sector principle were absolved from having to do more than chart the coastline to claim all the land that extended from there to the South Pole.

The sector principle particularly suited the British Empire, which had applied it in the Arctic to the lands and territorial seas within lines of longitude from Canada to the North Pole, while the United States could only claim the smaller wedge north of Alaska. Applying the same principle in Antarctica also suited Britain, whose explorers had charted much of the coastline and could therefore claim the huge sectors that extended to the South Pole. But these claims depended upon the sector principle being universally recognized. The United States, unsurprisingly, opposed the sector principle in the Arctic because it allowed the British Empire to maintain a chokehold on the much-sought-after Northwest Passage linking the North

Atlantic and Pacific Oceans. Thus the United States didn't recognize the sector principle in Antarctica. Yet Byrd implicitly accepted the sector principle during his expeditions in the 1920s and 1930s, when he flew across huge swathes of the continent to the east of Britain's Ross Dependency and declared that his exploration of Marie Byrd Land had made all that territory from the coast to the South Pole a possession of the United States. His claim still required an explicit declaration to that effect by Washington, and in the 1930s the American government was not yet prepared to do so.

Norway was similarly averse to using the sector principle in the Arctic, since it could impinge on Norwegian whaling activities in the Arctic Ocean. It wasn't only the Canadian sector that concerned the Norwegians but also the large sector of the Arctic that could be claimed by Denmark because of its control of Greenland. So, Norway made clear that its territorial claims in the Antarctic would only encompass the coastline and the immediate hinterland. This prevented London from agreeing to a deal with Oslo whereby they would each recognize the other's territorial claims. With Norway refusing to recognize Britain's sectors, and Britain refusing to recognize any of the more limited claims that Norway might make, the two nations simply settled on not actively opposing the other's claims.

While Britain could count on Australia and New Zealand to accept the sector principle, it was no help for Britain that Argentina and Chile also accepted the sector principle in Antarctica, since their territorial claims were much stronger than Britain's and extended all the way from the South American continent to the South Pole. And then there was France, which had claimed Adélie Land by the raising of the French Tricolour on a small offshore island more than eighty years earlier. France was adamant that the whole territory all the way to the South Pole remained French despite it never having been visited again in the intervening years by another French citizen.

How did the use of aircraft affect the claiming of territory?

There was some doubt as to whether merely seeing a place from a passing ship was sufficient to give any rights to its ownership. Dumont d'Urville certainly thought he had to do more, believing that it was important to stand on the territory he was intent on claiming rather than claiming it from the deck of his ship. A century later, aircraft transformed Antarctic exploration and raised new questions as to what was required to claim a newly discovered land. An aviator could bring more of the continent within his view on a single flight than earlier explorers could do in a whole season of dog sledging. When they flew beyond the coastline, these aviators were usually the first humans to have seen the "land" over which they were flying. Those flights of discovery were especially important for nations like the United States, which as noted did not adhere to the sector principle and argued that all the territory that was first flown over by its aviators should rightfully be theirs.

Richard Byrd thought that he could claim for the United States ownership to all the snow-covered territory that he could see from his cockpit, estimating that it stretched for 150 miles on each side of his flight path and for the same distance beyond the furthest point of his flight. At the same time, he was also conscious that much of the territory had been explored long before by the Scott and Shackleton expeditions, as well as by Amundsen, and that the British and Norwegians had claimed it. Although Byrd had promised to respect those claims by raising a British and Norwegian flag when he reached the South Pole, he didn't do so. Perhaps it was because he didn't land at the South Pole, as he originally intended, but simply flew over it. However, as we've seen he did drop an American flag from the aircraft that had a stone from the grave of the co-pilot on his flight to the North Pole wrapped in it.

Byrd was hedging his bets. Although the dropping of the flag on the South Pole could be used to buttress an American claim to Marie Byrd Land based on the sector principle, he was conscious that he hadn't landed at the South Pole and had dropped a flag rather than raised it. Byrd hoped that his maps of the region, based largely upon aerial photography, along with the American names that he'd given the geographical features would support the territorial claim. However, aerial mapping was still in its infancy, and it proved impossible to translate Byrd's spectacular photographs into accurate maps. In the event, the maps didn't matter, since Washington refrained from making an official claim to any part of Antarctica.

Lincoln Ellsworth was also captivated by Antarctica in the 1930s. He was the heir to a coal-mining fortune and was motivated more by a spirit of adventure and love of the wilderness. Rather than establishing bases, Ellsworth planned a historic flight across Antarctica from the Ross Sea to the Weddell Sea in 1934. However, it came to nothing when his plane was left too close to the ice edge and drifted out to sea. Returning the following year, Ellsworth planned to fly instead from Deception Island to Little America only to be defeated by bad weather. The aging adventurer returned in November 1935 and this time succeeded in flying the length of the peninsula before being compelled to land and walk the last 160 kilometers to Little America after his plane ran out of fuel. He'd claimed territory for the United States and named it after his father.

How did the United States come to dominate Antarctic exploration during the interwar period?

The exploration of Antarctica in the interwar period was being increasingly dominated by the United States. Byrd hustled up the finances, generated the popular interest, and mustered the all-important political support for his large-scale expeditions. But it wasn't Byrd alone. He was only able to mount his expeditions because of support from powerful business interests. Aircraft manufacturers were keen to offer their aircraft for use in the Antarctic. It was a time when aircraft had an increasing grip on transporting people and goods, but there was still widespread nervousness about flying. Byrd's flights helped to allay those concerns and provided invaluable advertisements for aircraft manufacturers. His expeditions were also a gift for newspaper publishers, while radio stations broadcast the latest bulletins from the expedition and cinemas delighted moviegoers with films of the Antarctic scenery and wildlife.

There was money to be made from the Antarctic, and the American media companies had the biggest audiences and budgets, which they used to support the activities of Byrd and his rival, Lincoln Ellsworth. The American dominance during the 1930s was also due to the support that Byrd received from the U.S. government, with President Roosevelt being keen to extend American power into the Pacific to confront the rising power of Japan and China. In comparison, the British expedition of John Rymill from 1934 to 1937, because it was mainly scientific, generated little media interest and had to operate on a shoestring. Although the Norwegian whaler Lars Christensen supported exploration to protect his commercial interests, and his whaling ships provided logistical support for his expedition ship, it was a small-scale operation compared to that of Byrd. Then again, although Byrd and the Americans dominated Antarctic exploration in the 1930s, his two ships, several aircraft, and seventy or so men were vastly outnumbered by the fast-growing whaling fleets, which comprised hundreds of ships and thousands of personnel.

Why was there a resurgence in the hunting of whales during the interwar period?

A sharp rise in the price of whale oil and the use of factory ships from 1925 onward allowed whalers to rove much more widely across the Southern Ocean. The installation of rear slipways in the late 1920s allowed whales to be hauled aboard by their tails and flensed on deck. The main danger was roping the tails in pitching seas, but this was soon solved with the introduction in 1932 of mechanical claws. Other technological innovations saw the boilers moved below deck, although the process of boiling remained a tough and dangerous job with workweeks of seventy hours or more.

The interwar period also saw oil replacing coal as the fuel for whaling ships. Tankers accompanied whaling fleets to refuel the ships and thereby allowed them an expanded range. Once the tankers were empty of their fuel, the tanks could be cleaned and whale oil pumped aboard. While that was good for the whalers, it wasn't good for the whales. Tens of thousands were killed each year during the 1930s, mainly blue, fin, and sei whales, with each species becoming progressively scarcer in turn. Although whales had a brief reprieve when the value of their oil plummeted at the beginning of the Depression in the 1930s, a gradual recovery of prices saw the fleets again head south and the killing resume with even greater intensity.

Why were nations unable to agree on a sustainable way of harvesting whales?

The frenzy of killing that characterized the sealing trade in the early nineteenth century was replicated by whalers in the 1920s and 1930s. And the results seemed set to be the same, with the likely extinction of most whale species from the Southern Ocean. Although environmentalists had been calling since 1910 for whale sanctuaries to be created, their appeals were ignored. Rather than being curtailed, the killing in the Southern Ocean increased from 11,000 whales in 1919–1920 to more than 40,000 a year by 1928–1929. In 1931, there were nearly three hundred whaling ships operating in the Antarctic. The blue whales were the first to disappear. Because of their enormous size and the amount of their blubber, they were targeted by whalers in the early 1930s until there were few to be found. Rather than putting their killing on a sustainable basis, the whalers simply moved onto other species. Such an approach was clearly unsustainable.

The British government controlled the Falkland Islands Dependencies. With its naval and economic power, it was best placed to bring the killing under control. Britain might have been successful had it acted quickly. By licensing whalers to hunt in Southern Ocean waters over which Britain claimed ownership, the British government could impose restrictions about which whales could be killed and require the whole carcass to be processed rather than just the blubber. However, British sovereignty could only be asserted within three miles of land. The position of the coastline was often impossible to ascertain in the Antarctic, when an ice shelf of uncertain length extended from the land into the sea, sometimes for many miles. This was particularly problematic in the Ross Sea, which was popular with whalers and administered by New Zealand. The uncertainty about their distance from land allowed unlicensed whalers to declare that they were hunting in international waters.

In the summer of 1926–1927, a Norwegian whaling fleet was operating in the Ross Sea, with a New Zealand official aboard to check that they were abiding by the conditions of their expensive license. The law-abiding Norwegians were dismayed to encounter other Norwegian ships working without licenses and ignoring the limited conservation measures that Britain was attempting to enforce. There was little that Britain could do about enforcing those measures, unless whaling nations agreed to an international treaty to make the industry sustainable. However, Britain was reluctant to call an international conference for fear of restricting the revenue of its own whalers and allowing the Norwegians to get a guaranteed quota of whales.

Enforcing such a treaty became much more difficult from the mid-1930s, as Japan and Germany began to dispatch whaling ships to the Southern Ocean. By 1938, Nazi Germany was sending five factory ships and fifty fast whale catchers with harpoon guns on their bows. With Japan already embroiled in a war with China, and Germany intent on launching a war in Europe, they needed whale oil for their fast-expanding war machines as well as to feed their populations. Neither nation was receptive to the conservationist arguments of the British government, which were seen as self-serving attempts to protect the British and Norwegian whalers from their new competitors. Fortunately for the whales, if not for the world, the slaughter came to a temporary end with the outbreak of war in Europe in 1939 and in the Pacific in 1941. But the war did not end the imperial rivalry for control of Antarctica, which continued throughout the conflict, albeit more surreptitiously and with fewer players.

4

WAR ON THE ICE

Why did Nazi Germany send an expedition to Antarctica?

Although Germany had been involved in Antarctic exploration since the early twentieth century, its expeditions failed to meet the expectations of their supporters. In the 1930s the Germans also had a growing involvement in the whaling industry, sometimes being the effective owners of whaling companies that were ostensibly Norwegian. They had a vital interest in whale oil, both as an essential foodstuff and as an ingredient in manufacturing explosives. By the late 1930s, a growing German whaling fleet sailed each spring to the Southern Ocean. Like the Norwegians, they were confronted by a continent that was unoccupied but nevertheless overlaid with the territorial claims of other nations. The Nazi government decided to make a claim of its own.

Germany might have been forced in 1919 to give up any rights it had in Antarctica, but there was no reason why they now couldn't make new discoveries upon which to base fresh territorial claims. An expedition could support the work of the German whaling fleet and bring prestige to the nation. It could also gather intelligence about the possibility of isolated harbors that its warships could use in the event of another war. It was with these aims that the expedition ship *Schwabenland* steamed south in the summer of 1938–1939. There were no dogs or sledges aboard, since the Germans were intent on proving their technological superiority to the explorers of yes-teryear. This was not an older whaling or sealing vessel, but a specially designed ship that could catapult aircraft into the air and then retrieve them on board with a crane. It meant the Germans didn't have to await calm seas before launching their flights. The expedition also had specially designed javelins to drop onto the ice from their aircraft. The heavy javelins were complete with swastikas and were meant to plunge into the ice with the point downward and the swastika standing proudly upright. Thus the expedition would avoid having to drop a cloth flag onto the ice, as other explorers had done, only to have it covered by the next blizzard.

Despite its ambitions, the expedition suffered the same deficiencies as other explorers who relied upon aircraft to pho-tograph and map the continent. Without ground control points to judge distance and the height of geographical features, no accurate maps could be produced from photographs alone. That didn't prevent the Germans from issuing grandiose announcements about their achievements, along with photos of several unfortunate penguins that were brought back for the Berlin zoo. Because of the war, the follow-up expedition planned for the summer of 1939–1940 never eventuated.

What was the reasoning behind the United States Antarctic Service Expedition of 1939?

The news of the German expedition prompted President Franklin Delano Roosevelt to send America's first official ex-pedition to Antarctica for more than a century. With Japan and Nazi Germany sending ever-larger whaling fleets, and Germany also sending an expedition, Roosevelt decided that the United States could no longer hold back. He had previously encouraged private expeditions and gave them a semblance of official blessing by issuing commemorative stamps. In 1939, Roosevelt went much further by appointing Admiral Richard

E. Byrd to lead an official American expedition. Dubbed the United States Antarctic Service Expedition, it was meant to signal the beginning of a permanent American presence on the continent. Americans would do what other nations had been unable to do: establish a permanent base. In fact, the base was to be occupied only during the summertime, rather than have personnel remain throughout the debilitating winter darkness. Nevertheless, it was intended to do more than any other nation had ever done.

The Americans were also going to outdo their rivals by mapping the entire continent. The means of achieving this grandiose ambition was a so-called snow cruiser, a motorized vehicle that was supposed to be able to surmount all the obstacles that usually confront motorized vehicles in a crevasse-ridden terrain. The snow cruiser was the size of a railway carriage and was fitted with the largest tires ever built. Each wheel had a separate motor that provided for each wheel to be lifted clear if it was stopped by an obstruction or slipped into a crevasse. As well, the snow cruiser was designed to move at a top speed of almost 30 miles an hour and had a small aircraft on its roof. The aircraft allowed the snow cruiser to have an even greater range, and with these capabilities it was predicted to be able to map and photograph the entire continent. As a mark of these ambitions for the snow cruiser, Byrd took with him thousands of stamped envelopes embossed with the words, "The Snow Cruiser Reaches the South Pole."

The other part of the American plan was to replicate the proclamations and flag-flying (or flag-dropping) that other nations had done. The U.S. State Department provided Byrd with forms that could be placed in stone cairns or dropped from aircraft. The proclamations would note the name of the expedition, the place where the proclamation was left, and the date when it was done. Each claim was carefully recorded and later given to the State Department. This activity was to be kept strictly secret until Washington was ready to make its claims known to the world.

Byrd intended to supplement the feats of the snow cruiser and its aircraft by making a flight of his own—one that would recreate his 1929 flight to the South Pole and continue until he had flown clear across the continent from the Ross Sea to the Weddell Sea. This flight was something that had never been done before. It would be made more magnificent by Byrd's establishing a ring of bases around the continent, thereby laying the basis for a possible American claim to the entire continent. It was part of Roosevelt's plan to extend the coverage of the Monroe Doctrine all the way to the South Pole.

Unfortunately, the grand plan fell to pieces. The snow cruiser had not been tested on ice, and it struggled to move in any way other than in reverse gear. Rather than roving across the continent, it became a stationary laboratory at Little America. The numerous bases that Byrd had planned were reduced to just two. The other one was established on Barry Island, where an abandoned British base was commandeered, and the island was renamed Stonington. The lack of a support base on the Weddell Sea precluded Byrd from achieving his flight across the continent. More crucially, the U.S. Congress refused to approve funding of the expedition for more than one year, which forced a hurried evacuation in 1941. Apart from the hostile Congress, the attack on Pearl Harbor ensured that continuation of the expedition was untenable. It proved to be a humiliating failure. There had been no worthwhile science, and most of the continent remained unseen and unphotographed. In the end, the American presence was as temporary as its predecessors. Even though Byrd told journalists that his expedition allowed America to claim 770,000 square miles (2 million square kilometers), the ownership of Antarctica remained up for grabs.

How was the idea of applying the Monroe Doctrine to Antarctica received by other nations?

Roosevelt's reference to the Monroe Doctrine to justify his establishment of the United States Antarctic Service Expedition caused alarm in Argentina and Chile. Both countries feared that the American move threatened their own claims to the areas they regarded as being integral parts of their own national territories. Their supposed entitlement went all the way back to the Treaty of Tordesillas, signed in June 1494, and based on a decision by Pope Alexander VI to divide up the world between the Portuguese and Spanish empires along a line of longitude that saw Portugal take Brazil while Spain took the remainder of South America. Byrd's 1939 expedition now put Chile and Argentina on notice about America's long-term ambitions in the Southern Ocean. Although the expedition had failed, it didn't mean they wouldn't be back.

Confident that Britain was distracted by the war in Europe, the Chileans and Argentinians informed London of their separate claims to Antarctica. Whereas Argentina asserted ownership of the area between 20° W and 68° W, Chile issued a decree in November 1940 claiming all the territory between 53° E and 90° W, which not only overlapped Argentina's claim but also that of Britain and the United States. The dispute might have been resolved at a planned conference to be held in Norway for just that purpose in 1940, but the German invasion of Norway put an end to that. With a postwar peace conference likely to include the issue of territorial rights in the Antarctic, there wasn't a moment to lose.

Argentina took the lead by beefing up its existing Antarctic presence. The official status of the longtime meteorological station on Laurie Island was declared to be an Argentine post office, while an Argentinian warship called at Deception Island so that its sailors could paint their nation's flag on the wall of the unoccupied whaling station. It also called at other islands within Britain's Falkland Islands Dependencies, where the Argentine flag was raised and signs were left to assert its claim to the region. Once the British became aware of these challenges, they sent a warship to paint over the Argentinian colors on Deception Island and perform symbolic acts of their own. Undaunted, the Argentinians returned to paint over the British colors.

Until Byrd's expedition and the subsequent actions of the Argentinians, the British had been relatively complacent about their territorial claims on the peninsula, which were based upon the history of discovery by their explorers and the mistaken belief that it was impossible to have permanent bases there. Although the American bases had been hurriedly evacuated in 1941, the expeditions by Byrd had shown clearly enough that permanent bases were feasible. That effectively demolished the British legal argument that its territorial claims could be sustained by occasional visits and flag-raising by passing expeditions. If the British wanted to hold onto their Antarctic territories, they would have to match the Americans. They would also need to ward off the increasingly assertive steps being taken by the Argentinians, whose base on Laurie Island was the only permanent base in the Antarctic.

As a first step, Britain sent a small party of marines in the summer of 1943–1944 to set up secret bases on Deception Island and at Port Lockroy on Wiencke Island. Plans for a third base at Hope Bay on the tip of the Antarctic Peninsula were abandoned due to thick ice that blocked access. On Deception Island, the marines repainted the British flag onto the wall of the whaling station before setting up inside. Britain could now claim that it was occupying, rather than just administering from afar, the territory it purported to own. Whether those few men occupying two bases would suffice to claim all the Falkland Islands Dependencies for Britain would only become clear once the war was over.

How was the historical record of exploration used to buttress territorial claims?

By the time of World War II, no nation had been able to establish an indisputable claim to any part of Antarctica. Yet even during the war, historians and geographers were enlisted to support the work of their nation's explorers by showing how their record of discovery was better than that of their rivals. Nations had to convince themselves, as well as their rivals, of their superiority in this regard. While explorers trudged through thick snow and carefully picked their way across hidden crevasses, historians leafed their way through dusty archives in search of a crumbling ship's log or fading map that would prove that their nation's sailors had been first to gaze upon the marvel that is Antarctica. The arguments went on, even as the war was raging, as to which nation had the stronger claim to different parts of Antarctica.

The American geographer William Hobbs devoted much of his later career to contesting British claims of discovery and promoting claims by his compatriots, especially that of Nathaniel Palmer. In the late 1930s and early 1940s, the arguments between Hobbs and his British counterparts were particularly fierce, as allegations about the falsifying of evidence were tossed back and forth across the Atlantic. At the same time, more measured contributions came from geographers at the Library of Congress, the U.S. State Department, and the American Geographical Society. But the aim of all of them was similar: to reinforce the primacy of American discoveries over those of their rivals.

It wasn't only old maps that were deployed in the prosecution of these arguments. The compilation of new maps was just as crucial. Because of this, the United States Antarctic Service was sustained for several years after the expedition's return in 1941 so that a small team of geographers could produce new maps with American names firmly affixed on them. The 1943 U.S. Navy publication *Sailing Directions for Antarctica* was the result. They hoped that Australian and British maps from the late 1930s might thereby be superseded. After all, the nation that could compile the most detailed map of the continent, complete with a multitude of place names bestowed by its explorers, would have a clear advantage. Once produced, the maps were then distributed as widely as possible in a vain attempt to have each nation's particular version accepted as the standard map of the continent. However, the greater the differences in place names between the various maps, the clearer it became that no standard map would be possible until agreement could be reached about the many competing place names. And that would take several more decades to achieve.

What was Operation Highjump?

The end of World War II saw the United States straddling the world, with its armed forces ensconced on every continent except for South America and Antarctica. Just as Britain had tried after World War I to bring all of Antarctica within its bloated empire, so too did the United States implement a similar plan after World War II. As the preeminent postwar power, the United States was the only nation able to mount a serious assault on the icebound continent. It was dubbed "Operation Highjump" and organized as a military operation. A naval task force of twelve ships was assembled, complete with the aircraft carrier USS *Philippine Sea* and the submarine USS *Sennet*. Nothing like it had ever been seen before in the Southern Ocean, and the veteran explorer Richard Byrd was in control.

It was Byrd who had pressed Washington to mount this expedition, using its ample armed forces and military equipment, so that America could take a commanding seat at any international conference to decide ownership of the continent. It wasn't just about the Antarctic. The United States' defense forces were preparing for a possible war against Russia and believed that much of the conflict would be conducted in the

Arctic, which provided the shortest route between the two nations. To fight a war in the Arctic, with Greenland as a likely base, would require much testing of American personnel and equipment in cold conditions. The Antarctic offered an arena where ships, aircraft, and tanks could be deployed in the world's coldest environment while being safe from Soviet surveillance. There was also a hope that Antarctica might contain deposits of uranium, which had become a vitally important strategic resource after the development of atomic bombs.

The submarine did not last long in Antarctica. As it edged its way into the Ross Sea in January 1947, it was gripped tight by sea ice and had to be extricated and sent home. The aircraft carrier was also not built to operate in pack ice and was sensibly kept well clear of it. The carrier had six DC-4 transport aircraft, which were flown onto the ice at the newly established Little America IV base. Like its predecessors, the base was situated on the Ross Ice Shelf. The planes were meant to do what the snow cruiser had been unable to do in 1940–1941. They would fan out across the continent, photographing its great expanse with newly developed cameras in the vain hope that a complete map might finally be possible. Such a map would feature American names, thereby maximizing the scope of any territorial claim that America might care to make. To support the claim, proclamations were to be dropped onto the ice while the aircrew displayed the Stars and Stripes on high.

Like Byrd's expedition in 1939–1941, the depositing of the proclamations was done in secret and in such a way that it did not commit the American government. Rather than being signed by President Harry S. Truman or some other top government official, the proclamations were signed by the men who placed them in a cairn or dropped them from an aircraft. Although the proclamations declared that they "claim this territory in the name of the United States of America," the wording allowed the State Department to argue that they did not constitute an official claim.

As so often happened in the Antarctic, the massive expedition over the first few months of 1947 did not go according to plan. The aircraft flew across 4 million square kilometers (1.5 million square miles) of the continent, much of it never seen before. However, some of it went unphotographed due to problems with the cameras and inadequate training of the photographers. There was also the familiar problem of the photographs not being supported by enough ground control points. Photographs of newly discovered mountain chains might look spectacular, but they are not useful to mapmakers unless their height and position can be accurately determined. As a result, the American dream of producing a complete and comprehensive map of the continent was thwarted again. Operation Highjump had been an expensive and ultimately futile exercise, with its scale and ambition only provoking other nations into planning expeditions of their own.

What were the effects of Operation Highjump?

The news of Highjump caused America's rivals great concern, since it threatened their own tentative hold on parts of the continent. Had the Americans' ambitions been achieved, they would have laid the basis for a claim to the entire continent. Even partly achieved, they posed a challenge to ownership of those territories over which American aircraft had flown or where the Americans had established their bases. The New Zealand claim to the Ross Dependency was under the most threat, since it was where Little America IV was located. As noted, Britain had passed responsibility for the administration of the Ross Dependency to New Zealand as a way of bringing most of Antarctica within the control of the British Empire. Although British expeditions to the Ross Sea had departed from New Zealand and some New Zealanders had accompanied those expeditions, that was not a strong basis for a territorial claim.

British lawyers now agreed that the mere discovery of a place counted for little and that raising a flag no longer sufficed to prove that a nation was its rightful owner. It required actual occupation. In late 1946, both New Zealand and Australia discussed whether to establish bases before the Americans could get there and prior to a possible conference on Antarctic governance. They even thought up some science that could be done as a pretext for their occupation. However, both nations lacked a vessel that could venture safely through the surrounding ice pack, and both were preoccupied with more pressing problems of postwar reconstruction at home. They were relieved when Operation Highjump came to an end without a permanent American base being established and the conference was postponed. In the absence of a base, Australia concentrated instead on updating its 1939 Antarctic map so that it would not be superseded by any map that America might make.

What was Operation Tabarin and its consequences?

While Australia, New Zealand, and France were unable to match the American challenge in the Ross Sea, there was a heightened burst of activity along the Antarctic Peninsula as Britain, Argentina, and Chile sought to preempt the Americans and ward off challenges from one another. Britain had established three bases during the war as part of the secret Operation Tabarin. As soon as the war was over, Britain quickly added two more bases. To give it a peacetime ring, Operation Tabarin was renamed the Falkland Islands Dependencies Survey, although its purpose remained that of demonstrating that Britain was in "effective occupation" of Graham Land and its offshore islands.

One of the British bases was on Laurie Island, where the Argentinians had maintained a base for more than forty years and where Argentina would otherwise have been able to make a claim to the whole South Orkneys. Another British base was set up at John Rymill's abandoned British base on Barry Island, which Byrd had taken over in 1940 and gave the island the American name Stonington Island. To emphasize Britain's sovereignty, each of its bases was designated as a post office and equipped with an official wireless station that broadcast weather information. Britain also issued Falkland Islands Dependencies postage stamps in 1946 as a further assertion of its sovereignty. Not to be outdone, Chile and Argentina responded in kind.

In the summer of 1946–1947, Chile sent a warship to the South Shetlands, where a small base, complete with a post office, was established on Greenwich Island (which was renamed President Aguirre Island). The Chilean flag was raised, the national anthem was sung, and a proclamation was buried beneath the building's foundation. To inform the world, two Chilean postage stamps were issued to celebrate its Antarctic territory, which extended from 53° W to 90° W. The Argentinians did likewise, sending warships to establish yet more island bases. The threat to the British territory became even greater when Chile and Argentina agreed in July 1947 to adjust the boundaries of their claims so they no longer overlapped, thereby setting the stage for a combined show-down with Britain.

At the same time, a semiofficial American expedition set off in 1947 under the command of the explorer Finn Ronne, a veteran of the Byrd expedition of 1940 and a U.S. Navy captain during the war, to occupy the base on Barry/Stonington Island. Ronne called on the British marines who were occupying the abandoned American buildings to vacate them, which they grudgingly did. Although he had the loan of a U.S. Navy tugboat as his expedition ship, Ronne made a pretense of acknowledging the Chilean claim to the area by having his passport stamped with a visa for the Chilean Antarctic and by taking a Chilean observer along with him.

Upon Ronne's arrival, the British marines retreated to their own buildings, where they flew the British flag and displayed a post office sign. When Ronne took possession of the American buildings and hoisted a Stars and Stripes, there was an immediate protest from the British, who wanted assurances that the flag did not signify an American claim to the territory. In fact, Ronne had already secretly made such a claim in 1940 when he was a member of the Byrd expedition, and he now stood by his right to again raise, as he declared to the British officer, "the American flag on the American-built flagpole at the American camp."[1] With neither side acknowledging the territorial claims of the other, with each side calling the island by a different name, and with the Chileans having their own claim to the region, the British and Americans nevertheless managed to cooperate in the business of exploring the much-disputed Antarctic Peninsula while keeping a sharp eye on their Chilean and Argentinian rivals. It was another chapter in the long saga of competition for control of the peninsula.

How serious was the risk of armed conflict in Antarctica at the start of the Cold War?

The environment has always been the greatest threat to life and limb in Antarctica. The extreme cold and wind of that frozen desert, as well as its hidden crevasses, had taken the lives of several explorers. Fire was also a constant danger that had to be guarded against. Once fire caught hold of a wooden hut, there was little possibility of extinguishing it in the absence of an ample water supply. In November 1948, the British base at Hope Bay on the Antarctic Peninsula caught fire despite the snow that had built up on its walls. The two men inside were unable to quell the flames or escape from the inferno, while the fire was spread by the wind to a store dump 200 meters (650 feet) away. In 1952, the recently built French base in Adélie Land caught fire and was partially destroyed, forcing the evacuation of the whole expedition and delaying France's Antarctic plans until a new base could be built.

The British had only been at Hope Bay for four years and were slow to return. That gave Argentina an opening. By the time the British returned in February 1952, they found two Argentinian supply ships at anchor and a newly constructed base on shore. Undaunted, the British began to unload their supplies, only to be confronted by the commander of the Argentinian base, who regarded Hope Bay as an integral part of Argentina and defended it by firing an automatic weapon over the heads of the British landing party. Although the British sensibly withdrew, a British naval frigate with a party of marines soon arrived from the Falkland Islands. This time, it was the Argentinians who sensibly withdrew in the face of the superior force.

Another such conflict could easily have erupted on Deception Island, where the British, Chileans, and Argentinians all had bases to buttress their competing claims. Tensions flared in January 1953 when an Argentinian supply ship arrived with ten tons of soil from Argentina. A naval party erected a hut and a flagpole close to the British base, only to have a Chilean party erect a symbolic hut of their own even closer to the British. The challenges to British sovereignty were answered with the dispatch of another frigate and a company of marines who dismantled both huts and arrested two Argentinian occupants as "illegal immigrants." It only inflamed the situation, as Argentina responded with air force flights over Deception Island and a warship tour in 1954 of all its Antarctic bases, accompanied by its naval minister.

The sword rattling could have gotten worse. Fortunately, cooler heads prevailed. Britain's empire was overstretched and its government had little appetite for armed conflict over territories that were not regarded as being of vital national importance. For their part, the Argentine and Chilean governments were happy to keep pressing their case and extending their presence while avoiding actions that would provoke outright conflict with the British or with each other. Despite this, the potential for war would remain as long as there was no international agreement to settle the competing territorial claims.

How were conflicts over place names resolved?

The naming of geographical features, such as mountains, bays, and glaciers, had a practical benefit for explorers, since it helped them determine their location and make sense of their maps. By covering the continent with the names of colleagues, benefactors, and family members, or other familiar names, explorers helped to make the forbidding landscape seem more welcoming and less alien. It also allowed nations to enjoy a greater sense of ownership over areas that their citizens had explored. All too often, though, multiple names were attached to the same place, as nations vied to be acknowledged as the rightful owner. Because of the Shackleton, Scott, and Amundsen expeditions, the polar plateau had both British and Norwegian names, with both nations purporting to be the owner. It was more complicated on Barry Island, which by 1951 had British, American, and Argentinian names.

The names applied on a map by one nation had little value unless other nations recognized them on their own maps. This was only done grudgingly, if at all, in those areas where there were not multiple claimants for the same territory. In the 1940s and 1950s, Britain tried to create a standard map of Antarctica that would privilege British names and discoveries. Britain had the advantage of being able to obtain ready agreement from Australia, New Zealand, and South Africa to the names it wanted to use within the Falkland Islands Dependencies. In return, it adopted Australian and New Zealand suggestions for their Antarctic territories. Norway was also willing to accept British names, since its territorial claims did not overlap those of Britain. The British map even included a few of the Russian names that Gottlieb von Bellingshausen had deployed in 1820. This was done on the understanding that Bellingshausen had not made a territorial claim at the time, and Russia had not done so since then.

A token scattering of old Russian names could not ensure that the British map would be accepted internationally. For that to happen, Britain needed the approval of Argentina, Chile, and the United States, which all had competing maps of the Antarctic Peninsula on which their names naturally predominated. No meaningful agreement could be reached on the naming issue until the various claimant nations were ready to set aside their rival territorial claims and develop a process for settling on a single name for each geographical feature. And there seemed little chance of that during most of the 1950s.

What part did uranium play in the drive to seize and occupy Antarctica?

The American destruction of two Japanese cities in 1945 alerted the world to the power that could be released by splitting the atom, which required uranium as its raw material. As Russia and the United States began their tense Cold War standoff in the late 1940s, there was a rush to secure supplies of the relatively rare and strategically important metal. Antarctica was promoted as having undiscovered deposits of uranium. In 1946, Australian Douglas Mawson suggested to reporters that Antarctica had reserves of uranium, though he had discovered none during his expeditions. But he argued that it must be present there because of Antarctica's resemblance to the Canadian Arctic, where uranium was then being mined. For Mawson, this was more about getting the Australian government to set up permanent bases in the Australian Antarctic Territory that he had done so much to explore and claim. The claims might also excite the interest of the public and make any government commitment appear justified. Mawson's predictions were one of the rationales when the government finally announced in 1953 that an expedition would be sent to the Australian Antarctic territory.

The Australians were not the only ones taking Geiger counters to Antarctica. The Americans and British had been doing so since the war, as they sought to find and secure deposits and keep them from the hands of the Russians, who had not yet developed an atomic bomb. Uranium was just one of the resources that interested the scientists and service personnel of Operation Highjump. Because of its strategic significance, the search for uranium was cloaked in secrecy, with Byrd even denying that America was in a race with Russia and Britain to find the mineral.

Of course, the thick layer of ice covering most of the continent prevented the discovery or exploitation of uranium beneath its surface. The greatest ice-free areas of the Antarctic were then as they are now on the Antarctic Peninsula and its offshore islands, which is where the British had their bases and were secretly examining rocks for any useful minerals. When Finn Ronne's expedition set up its base on Barry Island in 1947, the marines at the adjacent British base were instructed to conceal any sign that they were searching for minerals. In the event, nothing was found. The likely costs of exploiting mineral resources in the Antarctic have always been too exorbitant to consider. That also proved true with uranium.

How did the Cold War continue to play out in the Antarctic?

In 1947, the U.S. Central Intelligence Agency produced a secret map that purported to show all those scattered and convoluted parts of Antarctica that had been discovered by Americans, whether from the decks of their ships, from the cockpits of their planes, or on foot with their dog sledges. The map was drawn up to support a prospective claim to all those parts, whether or not they conflicted with the claims of other nations that were based upon the sector principle. The resulting mishmash of a map was kept secret and the territorial claims were never made. In the context of the Cold War, it was not worth the risk of rupturing relations with the important allies that were also America's territorial rivals in Antarctica—Britain, France, Norway, New Zealand, Australia, Argentina, and Chile. Russia remained the great enemy.

The United States had embarked on Operation Highjump in 1946 to train in the extreme cold of the Antarctic for a future war against Russia in the Arctic. For American military planners, Antarctica had the advantage of being safely out of sight of Soviet observers who might be looking on from ships or planes. Although several types of aircraft and naval ships, along with ground equipment such as tanks, were sent south for testing, the experiments didn't last long. By the late 1940s, American military planners had switched their attention to northern Greenland, where an air base was built to keep watch on Soviet movements and provide a front line in any future war. However, the United States still maintained a strong interest in the Antarctic and even considered making a formal claim, which would have created potential for conflict with the Soviet Union.

The possibility of the Cold War extending to Antarctica increased when Russia sent whaling ships there in 1946–1947 and its geographers reminded the world that the Russian explorer Bellingshausen was the first to discover Antarctica. There had been no Russians in Antarctica since Bellingshausen, but now Russian whaling ships were sharing the waters of the Southern Ocean with ships of the U.S. Navy. With the continent still having no widely recognized owners, the Russian presence raised obvious concerns in Western capitals about their possible intentions and increased the pressure on existing and aspiring claimants to establish permanent bases in Antarctica before the Russians could do so. That pressure became much greater when nations began planning scientific activities to be undertaken for the International Geophysical Year (IGY) in 1957–1958.

For several reasons, widespread fears of the Cold War extending to Antarctica were not borne out. The death of Joseph Stalin in 1953 helped to create a more cooperative relationship between the two superpowers. This was particularly true on a scientific level. Much to the surprise of the Americans, the Russian scientists involved with the IGY seemed only to be interested in the science and left their ideological baggage behind. Cooperation, rather than conflict, was also assisted by the Russians not using their armed forces to provide logistical support for their Antarctic effort. It might have turned out differently, as it did with the British and Argentinians, if both the Russians and the Americans had been armed and were operating in the same region. But the Russians were not armed and their bases were very distant from those of the Americans. The cooperation that developed between scientists working in the Antarctic raised the possibility of nations also being able to cooperate on the vexed issue of the continent's governance, which might thereby bring an end to long-running disputes over its ownership.

5

SCIENCE AND DISCOVERY

What role has science played in the discovery and fate of the Antarctic?

Science was a motivating factor in the dispatch of the earliest expeditions to Antarctica. In the eighteenth century, explorers and their sponsors wanted to fill in the empty spaces on the globe or to discover life forms that could be commercially exploited. By the nineteenth century, science became an end in itself. Explorers filled specimen jars with samples of life forms that were unique to the Antarctic region or gathered rocks that might explain the ancient geological history of the planet and its continents. Despite the interest in science, it was still money and power that provided the main motivations for governments and explorers. Robert Scott might have taken scientists on his two expeditions to the Ross Sea in the early twentieth century, but they were there merely to give the expedition an impression of seriousness. It was the same with many of Scott's contemporaries and successors.

Some of the early explorations, such as the Belgian expedition in 1898, the Scottish expedition in 1902, and the Rymill expedition in 1934, certainly prioritized science, even if the focus was primarily on geographical science. But it wasn't until after World War II that scientific discovery began to vie seriously with territorial claiming as the raison d'être for new expeditions. As early as 1949, scientists began to investigate the Antarctic for evidence of the apparent global warming that had been discerned in the Arctic. Glaciers were receding in the northern hemisphere, and scientists were keen to see whether that was also true in the southern hemisphere. Where better to look than Antarctica? That was one of the major aims of a combined Norwegian-Swedish-British expedition that left Oslo in November 1949 to establish a base on Dronning Maud Land, the large Antarctic territory to the south of Africa that had been claimed by Norway between the wars.

The expedition scientists were the first to drill ice cores in Antarctica, although their equipment did not allow them to drill very deeply. At the time, there was considerable disagreement about the depth of the polar ice cap. Seismic measurements made by Richard Byrd's expedition had found the thickest ice to be only 700 meters (2,300 feet) thick, but the Norwegians found to their surprise that their seismic detonations revealed the ice to be 2,400 meters (8,000 feet) thick. When they tried again they got the same results. They were also intrigued to discover various species of lichen covering the bare mountaintops some 300 kilometers (185 miles) from the coast. As one of the geologists busily collected rocks, he was taken aback to see a mite crawling on one of them. The mite was far beyond the reach of birds and other animal life that lived along the coast, and it seemed impossible for it to have survived such arduous conditions of intense cold and hurricane-force winds. Scientists have since surmised that these hardy mites feed on the lichen attached to rocks.

As for climate change, the scientists were unable to find any sign of the mountain glaciers retreating. Nevertheless, their findings helped to excite further scientific interest in Antarctica and added to calls for the International Geophysical Year (IGY) in 1957–1958 to focus on Antarctica in its worldwide scientific studies. In the planning for that event, science and scientists came into their own as Antarctic power-brokers, rather than merely acting as a cover for territorial ambitions.

When did scientists realize Antarctica was once part of a much larger continent?

The conventional explanation for the existence of similar plants and animals on different continents was that land bridges must have connected them in the ancient past before those bridges were submerged by rising sea levels. A land bridge linking South America and Antarctica was easy to imagine, given the relatively short distance between those two continents. But it was also believed that another land bridge connected Antarctica and Australia, which explained how marsupials came to exist in both Australia and the Americas and nowhere else and also why some plants, such as Proteaceae, are found in the fossils of Antarctica and present-day southern continents. During the nineteenth and early twentieth centuries, scientists searched for the bridge that had once supposedly connected Australia to Antarctica.

Another theory was that the continents had once been part of a single land mass that had broken apart and drifted over time to their present positions. However, the theory failed to gain general acceptance by scientists because a convincing explanation had not yet been found for the force that had driven the continents apart. It wasn't until the 1960s that the development of the theory of plate tectonics, the gathering of seismological evidence, and the examination of the deep ocean surface provided both the mechanism and the incontrovertible proof that explained the movement of the continents. It is now accepted that all the continents had been part of one supercontinent named Pangaea, which separated into two continents about 180 million years ago. After that separation, the southern continents of Antarctica, Australia, Africa, and South America, as well as Madagascar, Arabia, and India, made up a smaller supercontinent named Gondwana or Gondwanaland. Over millions of years, Gondwanaland moved apart along rift lines that eventually left Antarctica isolated at the bottom of the world, encircled by a cold current that created an ice-covered land. Rocks and fossils collected by a succession of explorers over the last two centuries have gradually filled in the picture of Antarctica's formation, while recent scientific discoveries suggest that the continent is still undergoing a slow process of separation that could see it split into two continents some millions of years hence.

Is all of Antarctica snow-covered?

Until the early twentieth century, explorers thought Antarctica was an ice- and snow-covered continent. While that's true for 98 percent of Antarctica, there are areas entirely free of such covering. Some of these clear places are mountaintops that have been blown clear of snow, but there are also extensive valleys largely devoid of ice. Robert Scott was the first explorer to stumble across one of three such valleys on the western side of McMurdo Sound in 1903. The vista made a spectacular contrast to the rest of the continent, with the exposed rocks of what became known as the Dry Valleys helping geologists to explore the formation of the continent.

The Dry Valleys cover more than 4,000 square kilometers (1,500 square miles). They were made possible by a mountain range that largely prevented the polar ice cap from pushing its way into them as well as by the dry winds that ensured evaporation of the little snow that falls on them. Yet the valleys have large glacial lakes formed by water melting from the glaciers above. The water from one glacier provides Antarctica with its longest river, which flows along 32 kilometers (20 miles) into Lake Vanda. The lakes of the Dry Valleys are totally ice-covered for most of the year, and if they ever melt it is only around the edges. While water on the surface of the 5-kilometer-long (3-mile-long) Lake Vanda is fresh, its depths of about 70 meters (230 feet) hold water that is surprisingly warm at more than 70° F and that is much saltier than seawater.

A smaller ice-free area was discovered in 1935 by the Norwegian whaler Klarius Mikkelsen and his wife Karoline, the first woman to set foot in Antarctica. Named the Vestfold Hills, they cover an area of 400 square kilometers (155 square miles) and, like the Dry Valleys, are kept free of ice and snow by a combination of the strong, dry winds and the hills that block the ever-moving polar ice cap. The area also has a multitude of lakes, most of which replicate the conditions found in Lake Vanda, with layers of water exhibiting a wide range of temperatures and degrees of salinity. Like Lake Vanda and the Dry Valleys, there are also microscopic life forms to be found in the lakes and among the rocks in this most inhospitable of environments.

Well before these lakes were discovered, the Russian explorer and scientist Peter Propotkin in the late nineteenth century proposed that lakes might exist deep under the polar ice cap. Propotkin thought that the immense weight of the ice would raise the temperature at the base to such an extent that the ice would melt and form lakes. A combination of airborne radar surveys in the 1970s and satellite readings in the early 1990s confirmed the existence of about 140 of these subglacial lakes across Antarctica. The largest of them, Lake Vostok, was discovered about 4 kilometers (2.5 miles) beneath the ice at the so-called Pole of Cold, where Russia had established a scientific station in 1957. It is one of the most massive lakes in the world, covering an area of about 12,500 square kilometers (5,000 square miles) and comprising more than 5,000 cubic kilometers (1,200 cubic miles) of water. In places, the lake is 900 meters (3,000 feet) deep.

Initially, it was believed that the water of Lake Vostok had been undisturbed for many millions of years, and some scientists thought it should remain that way. However, American scientists argued in 2002 that the upper waters of the lake were constantly freezing and being moved by the ice cap toward the sea, only to be replaced by water melting from other parts of the ice cap. Instead of the water having been there for millions of years, the scientists estimated that it was replaced by this process every 13,000 years. Some still argued that the lake should not be breached for fear of introducing contaminants into its pristine waters, which is what happened when Russian scientists drilled through the ice cap to reach the lake in 2012–2013. Using tons of chemicals to prevent the drill hole from freezing, the resulting water samples were duly contaminated and unable to show conclusively whether unique life forms had been able to exist in the lake's harsh environment.

Why have permanent bases been so hard to establish?

Setting up a tent in Antarctica was easy enough for the early explorers, but the establishment of permanent bases posed a much more difficult problem. It would have helped if an area of solid rock could be found on which a building could be erected. However, as Douglas Mawson discovered, even a wooden hut built on rock could be buried by wind-blown snow, forcing him and his colleagues to gain access by means of a roof hatch. At least the building would probably survive the onslaught, as Mawson's did, and remain available for future occupation. There was no such hope when constructing a building on the ice. Not only was it in danger of being buried by wind-blown snow, it also faced being slowly swallowed and then crushed by the ice. That's what happened to the buildings at the succession of Little America bases established by Richard Byrd on the Ross Ice Shelf. Each of the buildings sank inexorably below the surface of the slow-moving ice and were no longer visible by the time Byrd returned several years later.

The rush to establish bases during the IGY in 1958 was also a rush to discover ice-free areas on which the bases could be built. The Americans' familiarity with the Ross Sea, and its relative proximity to the South Pole, caused them to seize upon a site on the ice-free tip of Ross Island in the southernmost part of the Ross Sea. They also established the beginnings of a huge logistics base in February 1956 on a large expanse of land on McMurdo Sound, in territory claimed by New Zealand, and built the base next to Scott's hut near the edge of the Ross Ice Shelf. Its location allowed for easy off-loading of cargo and personnel by sea and the construction of an airfield on the nearby ice shelf capable of landing heavy cargo planes. McMurdo Station gradually grew to have the appearance of a frontier town, complete with a cinema and chapel and a criss-cross of streets. Between 1962 and 1972, McMurdo Station also had a small nuclear reactor, which made it unnecessary to transport huge quantities of diesel fuel for the generation of electricity. Such was the settlement's size that it vastly overshadowed the much smaller New Zealand base, named in honor of Scott, which was located just 2 kilometers (1.2 miles) away.

Australia rushed to reserve ice-free areas of its own along the coast of the Australian Antarctic Territory in the vain hope that this might keep the Russians out. The Vestfold Hills on the Davis Sea offered an extensive, ice-free area, comprising 400 square kilometers (155 square miles), that was adjacent to the coast of East Antarctica. It was there that Australia's Davis Station was established in 1957, with the Mawson Station having been built further west in 1954. The Russians were better resourced and in February 1956 they established their main logistics base, Mirny, between the two Australian bases, allowing them to use it as a jumping-off point for the inland Vostok Station. The Mirny base was built on four rocky outcrops on the shore of a small peninsula and survived the furious katabatic (downslope) winds that frequently sweep with cyclonic force off the polar plateau.

It was not just the wind that challenged the builders of Antarctic bases. One of the greatest threats to buildings and human life is fire. Because of conditions of an extremely dry atmosphere, winds, and a lack of water in Antarctica, a fire can quickly take hold and destroy a building. These conditions could have resulted in a catastrophe for the Borchgrevink expedition of 1898–1900, which nearly lost its living quarters when a mattress caught fire. This incident taught subsequent expeditions to guard against fire by erecting more than one building and dividing their stores among the buildings or burying them beneath the snow.

Even that precaution did not save the British base at Hope Bay in 1948 when the living quarters were burned down, killing two of the expeditioners, and gale-force winds sent sparks 200 meters (650 feet) away to a store dump. The sole survivor could only watch in horror after all his efforts at subduing the fire proved futile. Fortunately, he still had a tent in which to shelter and other expeditioners turned up within days to rescue him. The French almost experienced a similar disaster when their newly established base in Adélie Land caught fire in 1952, forcing its inhabitants to evacuate to a smaller French base about 60 miles away. Although a couple of small huts survived the inferno, the main base was never reoccupied.

Separating buildings, substituting steel for wood, and instituting strict protocols to guard against fire have not prevented other bases from being partially burned or even destroyed. A laboratory at the British research station on Adelaide Island caught fire from an electrical fault in 2001. Using a machine to blow snow onto the burning building could not extinguish the flames. There was no loss of life in that instance, but in October 2008 a Russian construction worker was killed when a two-story building at Russia's Progress Research station caught fire. Again, the blaze could not be extinguished although it didn't spread to other buildings. The personnel of a Brazilian base on King George Island were not so fortunate in February 2012, when a fire tore through its generator building before spreading to other buildings. Most of the base was destroyed and two men were killed. Although the base was evacuated, Brazil built a temporary replacement before constructing a new, larger base.

Apart from fire, bases built on ice must guard against being swallowed and crushed by it. Yet there is often no alternative for nations that are intent on siting their bases in a particular part of the continent, whether for scientific, strategic, or other reasons. The United States had compelling scientific and strategic reasons for establishing a base at the South Pole in 1956, where it was ideally situated for astronomical and cosmic ray research and also located in the best political place to assert American supremacy in Antarctica. The science could never have justified the expense, but the political rationale certainly did, even though the location made the expense all the greater because of the need to construct a new base each time the existing one was destroyed by the actions of ice and snow. The present base is the third to have been built on the site.

The original Amundsen-Scott South Pole Station was built by the U.S. Navy, using prefabricated wooden panels that were dropped by aircraft to waiting workers below. The buildings were always considered to be temporary due to snow drift accumulating against the wind-facing walls and eventually burying them. The structures could still be used, even when completely buried, and they were designed for that eventuality. By 1960, three years after their construction, these bases were under 2 meters (6.5 feet) of snow. And they continued to sink beneath the surface until they became too dangerous to occupy by 1975, when 15 meters (50 feet) of snow and ice covered them. A shift was made to a new, adjacent base. This one dealt with the snow drift by enveloping its wooden huts within a geodesic dome that was 50 meters (165 feet) wide and 16 meters (53 feet) high. The dome was covered with aluminum panels, which kept the wind and snow outside but could not prevent the accretion of icicles on the inner surface caused by condensation. The scientific "village" under the dome was designed to house about thirty personnel during the long winter and more than twice that during the summer season by utilizing tents. The dome cut off any view of the outside world, which was made more claustrophobic because most of the hut's windows were boarded up to conserve heat.

Despite the dome's size and imposing appearance, there was nothing to prevent snow drifts accumulating around the structure. Although the drift was regularly cleared away from the wall, the dome took on the appearance of an inverted silver basin sitting within an ever-deeper basin of snow. By the late 1990s, the dome was no longer usable. A new station was designed that could accommodate many more personnel and prevent the buildup of snow against its walls. The solution was to erect the metal-clad wooden structures on massive legs, with the two main buildings designed like an airfoil to force wind beneath them so that snow would be blown away. In the event of snow accumulating beneath the two-story buildings, the legs could be jacked up to keep them above the level of the snow. So much science was now being conducted at the South Pole that the buildings, which were completed in 2008, had to accommodate more than 150 personnel during the summer, making it one of the largest bases on the continent.

While the Americans are on the third incarnation of the Amundsen-Scott Station, the British have recently completed the sixth incarnation of the Halley Station, located on the massive Brunt Ice Shelf, which juts into the eastern stretches of the ice-choked Weddell Sea. Being on ice, the Halley Station similarly must deal with the problem of being swamped by snow drifts and slowly swallowed by the ice. The first incarnation of the Halley Station was completed in 1956 as part of Britain's contribution to the IGY. Neither it nor any of its subsequent incarnations lasted more than ten years before being buried deep beneath the snow and abandoned.

Halley and similar bases face an additional threat. Even if they could remain safely above the snow, building on an ice shelf that is moving inexorably out to sea means that the base will move out to sea as well. This is what happened to the fifth incarnation, the Halley V Station built in 1989, just 12 kilometers (about 7 miles) from the ice edge. Despite being on legs that could be jacked up, almost a meter (3 feet) of snow was accumulating beneath it each year. There was a limit to how high it could go to remain above the surface. That limit was not reached before there was a realization that the ice on which Halley V was built would calve off from the ice shelf and leave the base floating out to sea. Fortunately, this process usually happens over an extended time, which allowed the British to design a replacement before Halley V had to be evacuated.

The replacement base, Halley VI, was designed to provide a long-term solution to building on an ice shelf. The new base opened in 2012, and it was composed of several massive, interconnected pods held above the ice shelf on giant legs. The pods were shaped in a way to prevent the buildup of snowdrift. But its real innovation was the addition of giant skis on the legs that permit each pod to be moved if there is any danger of the ice shelf calving off. This is an ever-present danger with ice shelves, with global warming having made such calving events more frequent, and it caused even the ski-equipped Halley VI to be abandoned for the 2017 winter. An existing crack in the ice had forced the pods of Halley VI to be moved further inland during the summer of 2016–2017, only for a new crack to be discovered. It meant that the relocated base would not be safe for occupation over winter in case a calving occurred in the long months of darkness, during a time when it would have been impossible to rescue the fifteen personnel. As a result, the base was left empty during the winters of 2017 and 2018.

How was the hole in the ozone level over Antarctica discovered?

The benefits of doing science in Antarctica were graphically revealed in 1985 when investigators at Britain's Halley Station discovered a massive thinning of the Earth's so-called ozone layer above the South Pole. Ozone, a toxic but oxidizing gas, is largely created by the action of the sun in the tropics, from where it moves upward and outward driven by the wind until it reaches the stratosphere, about 10 to 50 kilometers (6 to 30 miles) above the earth. Consequently the stratosphere has a much larger percentage of ozone than the rest of the atmosphere. This concentration has the very beneficial effect of absorbing most of the harmful ultraviolet (UV) rays from the sun, which would otherwise threaten life on Earth.

Scientists have been studying the level and circulation of ozone in the atmosphere since the 1920s, with observatories around the globe taking regular measurements—by means of a spectrophotometer—to determine the amount of UV rays that reach the Earth's surface. By the late 1970s, scientists noticed that the amount of ozone in the stratosphere was beginning to decline. Overall, their measurements suggested a decline of about 4 percent in the ozone level, and this decline seemed to have no natural cause. Although scientists concluded that the declining ozone level must be caused by human action, there was no immediate alarm at the incremental depletion of ozone.

That all changed in 1985. Initially, scientists at Halley Station thought that their instrument must be malfunctioning when it returned apparently random readings at the low end of the scale. But there was nothing wrong with their instrument. Instead, a check of the results revealed that the ozone level was becoming dangerously low each spring, such that there was a virtual "hole" in the layer above Antarctica. More worrying, the scientists discovered that the hole was growing larger with each passing spring. If it continued, and the thinning of the ozone layer occurred elsewhere in the world, the implications for life on Earth were very serious. The increased UV rays reaching the Earth's surface would cause cancer in humans and damage the plants on which we rely for food.

It wasn't immediately clear which human activity was destroying the ozone layer. Further investigation quickly confirmed that it was mainly due to the production of chlorofluorocarbons (CFCs) for use in aerosol cans and as a coolant in refrigeration and air-conditioning units. Once released into the atmosphere, CFC molecules slowly rise into the stratosphere, where sunlight causes them to release their chlorine atoms, which then latch onto ozone and convert them into oxygen. It doesn't take much CFC to do great damage to the ozone layer, with one chlorine atom being able to convert thousands of ozone molecules into their constituent oxygen atoms. Once the culprit was identified and the chemical process understood, the world agreed in 1987 to the Montreal Protocol on Substances that Deplete the Ozone Layer (Montreal Protocol), which would gradually bring an end to the production of CFCs. Despite the Montreal Protocol, full recovery of the ozone layer is not expected for perhaps a century.

What surprising discoveries did scientists find beneath the ice cap?

For nearly two hundred years, scientists have sought evidence about the earliest beginnings of Antarctica and its geological composition. Fossil discoveries on islands off the Antarctic Peninsula in the early nineteenth century revealed that the continent had once been lush with vegetation and enjoyed a temperate climate. A century later, the early explorers looked for rocks that might indicate the presence of valuable minerals and suggest what lay beneath the ice cap, the depth of which was still unknown. Some indication of the continent's composition came from the 1872–1876 HMS *Challenger* expedition, which dredged the floor of the Southern Ocean for rocks that had dropped from melting icebergs. These results confirmed the existence of a continent that was similar in composition to all the others.

For nearly a century, this remained the best way to determine the geological composition of the continent. Even when the early explorers headed inland, it was difficult for them to extrapolate from the few rocky formations, called "nunataks," that could be seen and sampled. They could hardly imagine that these rocky outcrops were actually the peaks of mountains that can be 3,000 meters (10,000 feet) or more above sea level. That startling knowledge only came when scientists used a variety of instruments to look beneath the ice, allowing them to measure its thickness and gauge the nature and shape of the underlying continent. Richard Byrd made the first seismic investigations of Antarctica during his second expedition in 1933–1935. Because his investigations were limited, so were his results. On his later expeditions, Byrd supplemented the seismic investigations with the use of an airborne magnetometer in an attempt to reveal geological anomalies beneath the ice.

These early attempts hardly scratched the surface. It wasn't until the IGY in 1957–1958 that a large-scale, multinational attempt was made to uncover what had lain hidden for so many millions of years. To assert its own position in Antarctica, Britain mounted the Commonwealth Trans-Antarctic Expedition in association with New Zealand, Australia, and South Africa. Under the leadership of Dr. Vivian Fuchs, this expedition preempted some of its rivals by starting in 1955 and undertaking a feat designed to captivate public attention. The expedition took up the challenge that Ernest Shackleton had failed to accomplish in 1919. Using an assortment of snow vehicles and tractors, and supported by light aircraft, Fuchs left from a British base on the Weddell Sea and headed for the Ross Sea by way of the South Pole. His expedition was assisted by having another party, under the leadership of the famous mountaineer Sir Edmund Hillary, head off from a New Zealand base on the Ross Sea. The expedition established a line of supply dumps as far as they could get toward the South Pole.

The headline-grabbing feat was given a veneer of science by each party doing a seismic survey along the journey. It turned into a race when Hillary saw a chance to reach the South Pole before Fuchs's Weddell Sea party. It was like Scott and Amundsen all over again, with Hillary arriving at the South Pole on January 4, 1958, with barely any fuel left, beating Fuchs by sixteen days. It remained for Fuchs to claim the honor of being the first to cross the continent on land, albeit by mechanized transport, completing 3,475 kilometers (2,159 miles) in just over three months. As so often in Antarctica, science had become hostage to personal and national ambition. Although Fuchs's expedition ended in acrimony and was always more about derring-do and national aggrandizement than science, the results that Fuchs brought back were nonetheless important. Unlike Hillary, Fuchs wasn't following in anyone's footsteps during his historic journey from the Weddell Sea to the South Pole. His seismic measurements showed conclusively that the long-postulated frozen channel separating West and East Antarctica did not exist.

Many of the other fifty or so IGY bases were located on the Antarctic Peninsula or nearby islands, while the Americans and Russians established several bases on the polar ice cap. The American base at the South Pole and the temporary Russian base at the pole of inaccessibility produced some of the most important early results about what lay below the ice. Specifically, the Americans found nearly 3 kilometers (2 miles) of ice below their base, which meant the ice cap was much deeper than Byrd had suggested in the 1930s. The small rocky outcrops that occasionally poked above the ice became much more significant. The further that scientists looked below the ice, the more mountains they found. Rather than a channel bisecting West and East Antarctica, seismic investigations found a mighty mountain range stretching across the continent, with only some of the peaks of the Trans-Antarctic Mountains appearing above the ice. Although some traverses were done during the IGY to get a partial glimpse of the topography far below, it would take the introduction of a new technology in the 1980s to fully attempt a comprehensive topographical map of the whole landscape beneath the ice cap.

What impact did the advent of satellites have on Antarctic science?

Ever since the 1920s, explorers flying over Antarctica have tried to solve some of its biggest questions, such as whether it comprises a single continent or consists of two large islands divided by a channel between West and East Antarctica. For a long time, it was even suggested that there might be a vast inland sea at the South Pole. Byrd tried and failed to traverse the entire continent in 1939, and he failed again after World War II, when six long-range aircraft were used to criss-cross the continent in an unsuccessful attempt to map it for the first time. It wasn't until a satellite passed over the South Pole in 1964 that the first complete image of the continent could be taken.

The results were initially rudimentary. The first photographs were obscured by clouds and snow, and the outline of the continent could not be discerned with any accuracy because it was impossible to make out the coastline when it was edged with ice shelves and sea ice. With the launching of more sophisticated satellites over succeeding decades, radar and other equipment could see through the cloud cover and even differentiate between ice-covered land and a thick ice shelf, thus making possible more accurate maps. Satellites have recently been deployed to track the increasing amount of carbon dioxide in the atmosphere. The results have revealed the source of the gas and where it ends up, with about 30 percent being absorbed by the action of the ocean. The satellites have been able to show that the cold and stormy Southern Ocean has absorbed more than its share of carbon dioxide, with much of it sinking into the ocean depths. As carbon dioxide increases, it will pose a future threat to the krill population, and thus to the whole Antarctic ecosystem.

Satellites have been most important in tracking the effects of global warming. For one thing, they have documented the dramatic diminution of Antarctica's winter sea ice that has occurred over recent decades. But they have done much more. A joint American and German program, involving two satellites launched in 2002, measured microscopic changes in the gravitational pull of the geographical features over which they were passing. Known as the Gravity Recovery and Climate Experiment (GRACE), the satellites were designed to orbit the Earth just 220 kilometers (136 miles) apart, constantly bouncing microwave pulses between them as they measured the tiny changes in speed that were caused by the pull of the Earth below them. Minute variations in the distance between the two satellites were then correlated with the features over which they happened to be passing at the time. This simple idea, albeit requiring incredible scientific and technical proficiency, provided scientists with a wealth of information about the movement of water around the world. Among other things, they were able to measure the depletion and replenishment of the world's underground water supplies, which can be crucial to the lives of billions of people who depend on them. They were also able to measure the depletion of Antarctic ice shelves and the rate at which ice is being lost from the Greenland and Antarctic ice caps.

Whereas the early satellites could measure the area of the ice shelves and see whether they were shrinking or expanding, the GRACE satellites also measure the depth of both the ice shelves and the ice cap. This has allowed scientists to determine whether the total quantity of ice is increasing or decreasing. They found that the previous estimates of ice loss were much too low for both Greenland and Antarctica and that the rate of loss seems to be increasing for both places. This information is crucial for understanding how much water these two ice caps are contributing to the ongoing rise in sea levels. It is also useful in predicting how the rise in sea levels will play out in the future, as the disappearing ice shelves allow glaciers to accelerate toward the sea.

The data from the GRACE program have also allowed scientists to look beneath the polar ice cap and discover physical features that were hitherto unknown. In 2006, scientists examined satellite data of gravitational changes in Wilkes Land and noticed a strange anomaly in the data received from beneath the ice. It led them to suggest that there had been a huge asteroid, about 240 kilometers (150 miles) across, which had crashed into Earth millions of years ago and created a massive crater. Scientists have postulated that this asteroid may have been responsible for an even greater mass extinction event than the later asteroid that is widely believed to have been responsible for the extinction of the dinosaurs.

Starting in the late 1980s, communication satellites have also brought Antarctica much closer to the rest of the world. Until then, people working at Antarctic stations were largely isolated, with only intermittent wireless communication with their home countries and with other Antarctic stations. Teleprinter and facsimile messages were made by radio in the 1960s, as were radio telephone calls. The latter were expensive and subject to atmospheric interference, and they didn't do much to break the winter isolation. That all changed with the coming of communication satellites and the development of the internet, which allowed people to email their colleagues and their families as frequently as they required. The advent of social media, Skype, and other technology has brought the world even closer to Antarctica. Webcams now allow people from Beijing to Boston to log in to watch penguins in their rookeries and scientists in their labs. Communication satellites have allowed scientists to communicate their observations instantly and without interruption, whether it concerns cosmic ray activity or meteorology.

As scientific activity expanded in the decades after World War II, there was a concomitant decrease in the commercial activity that had been expected to drive future human involvement with the continent. During the first half of the twentieth century, there had been predictions about resources being concealed beneath Antarctica's ice cap or under its offshore waters—everything from coal to uranium and oil. Nothing had come of those hopes. After the war, there was only whaling, which had been the most profitable activity and had helped drive exploration prior to the war. But even the whalers proved to be a disappointment and never reappeared in the large numbers that had been seen in the 1930s.

6

COMMERCIAL ACTIVITY
IN POSTWAR ANTARCTICA

Why did postwar whaling peter out so quickly?

The lack of Antarctic whaling activity during World War II allowed whale stocks to begin recovering. But it didn't take long for whaling ships to reappear once the war ended in 1945. The British and Norwegians were quick to get back into whaling, though other nations were also eager to profit from what was still a lucrative industry and to employ whaling to secure foodstuffs for peoples devastated by war. Even the Dutch sent a factory ship to the Antarctic shortly after liberation of the Netherlands from the Nazis in 1945. A more significant newcomer was a Russian whaling fleet, which employed Norwegian whalers and used whaling ships seized from the Germans after the war. The presence of a Russian factory ship and associated whale catchers in Antarctica during the summer of 1946–1947 provoked concerns about Russian intentions, stoking fears that it could lead Russia to make territorial claims on strategic parts of the continent. In fact, the war-wracked Russians had a more prosaic plan to exploit the food and oil that whales could provide for the recovering Russian economy.

The Americans were motivated by a similar ambition when they allowed the defeated Japanese to deploy some of their surviving whaling ships to the Antarctic. The head of the American occupation forces in Japan, General Douglas MacArthur, wanted the fleet to provide whale meat for the half-starved Japanese population, as well as oil that could be sold on the international market to recoup some of the expense of the American occupation. The return of the Japanese to the Southern Ocean alarmed the Australians and New Zealanders, who had just helped to defeat Japan in the South Pacific and feared the return of Japanese ships so soon after the war. Australia also had its own whaling ambitions, which were threatened by the Japanese activity. In the event, the Australian whaling companies never went beyond their own coastal waters.

Unlike the unprecedented scale of the prewar slaughter, a measure of control was now imposed to make the whaling industry sustainable. The creation of the International Whaling Commission (IWC) in 1946 recognized that the northern hemisphere had been largely depleted of whales and that stocks in the southern hemisphere were in danger of disappearing if whaling nations didn't regulate the size and nature of the kill. The nineteen nations of the IWC agreed that the equivalent of only sixteen thousand blue whales could be killed in the first season; that calves had to be spared; and that the entire whale carcass had to be used, rather than just taking the oil-rich blubber and leaving the rest to rot slowly in the freezing water. It also decreed that only those nations involved in whaling prior to the war could become involved after the war. Those restrictions still saw fifteen factory ships and about seventy whale catchers head for the Southern Ocean in the summer of 1946–1947.

Because of postwar food shortages, whale oil was fetching prices far higher than those of the 1930s, which attracted even more whaling ships. In January 1951, the waters around Antarctica were being churned by 19 factory ships and 239 whale catchers. However, the killing was halted as soon as the annual quota was reached. Although research on which the quotas were based was still flimsy, it meant that the unbridled slaughter of the prewar years was not repeated. But the toll was still too much. Research on the impact for the whale population during the 1960s caused the IWC to ban the killing of blue and humpback whales altogether. Additional species were added to the protected list during the following decade, and all commercial whaling was banned in Antarctic waters from 1986 onward. Only "scientific whaling" was allowed thereafter. This condition provided a sufficiently large loophole for Japan to resuscitate Antarctic whaling (albeit on a much smaller scale) under the guise that it was conducted for scientific purposes.

Why did Japan resume whaling in Antarctica and why was the world unable to stop it from doing so?

The IWC monitored whale populations and set annual limits on the numbers that could be killed in different oceans of the world. By the 1970s, the IWC's members were under intense pressure from the environmental movement to bring the hunting to an end. With some species on the verge of extinction, the IWC finally agreed in 1982 that all commercial whaling had to cease by 1986. The commission's edict only allowed whaling by indigenous people for their own use and by nations for scientific purposes. Norway and Iceland refused to abide by the edict and have continued to hunt for whales in the North Atlantic, while Japan issued permits that allowed killing by Japanese whalers in the Antarctic, ostensibly for scientific purposes. Those permits continued to be issued despite the creation in 1994 of a whale sanctuary in the Southern Ocean that consisted of 31 million square kilometers (12 million square miles).

Japan has been adamant about its right to hunt for whales and continued to do so in the face of strong campaigns by antiwhaling groups, which included sending ships to the Antarctic to interfere with the whaling. First among these groups was Greenpeace, followed by Sea Shepherd. These antiwhaling groups took direct action against the whalers as they attempted to come between them and their quarries or to prevent refueling of the whale catchers or loading of dead whales on board factory ships. This high-stakes strategy caused several collisions between the ships and the sinking of one of the Sea Shepherd vessels. The groups' opposition strengthened the Japanese resolve to continue hunting minke whales in the Antarctic. As justification, Japan pointed to its long history of whaling and its use of whale meat as food for its people. It also pointed to the abundance of minke whales along with scientific evidence that a limited cull of their large numbers would not imperil the species and could even allow other bigger species to recover their own numbers more quickly.

These justifications don't tell the whole story behind Japan's determination to persist with whaling. There's also the affront that would be caused to Japan's dignity if it backed down in the face of international pressure. Any concessions would provide encouragement to environmental groups, who similarly object to Japanese fishing practices that threaten the survival of species such as Pacific tuna. With fish comprising such a large part of the Japanese diet, the government in Tokyo felt unable to compromise. Despite a ruling against Japan's scientific whaling by the International Court of Justice in 2014, the whaling and clashes with the Sea Shepherd ships continued. In the summer of 2016–2017 another 333 minke whales were taken. That activity looks set to cease now that Japan has announced its intention to withdraw from the IWC and allow the resumption of commercial whaling, which is likely to occur in waters close to Japan rather in the Antarctic.

What is krill and why did nations expect to benefit from harvesting it?

What draws tens of thousands of whales back to Antarctic waters each summer is a tiny, red-tinged, prawn-like crustacean that lives in the world's oceans, feeding on microscopic plant life. Antarctic krill only grow to about 6 centimeters long (about 2 inches) and weigh up to 2 grams (less than a tenth of an ounce). Yet they are so abundant, and swarm in such numbers in the Southern Ocean, that whales migrate thousands of miles to sieve the rich food source through their massive mouths. Feeding on the krill allows the whales and their calves to put on sufficient weight, much of it as blubber, so that they can swim the great distance back to their winter breeding grounds in the tropics. The krill also provide the major food source for penguins and comprise part of the diet for seals and fish.

The early whalers had occasionally witnessed the surface of the Southern Ocean turning red as swarms of krill erupted to the surface. It was a relatively rare event, as the krill mostly lived in the deep and usually came to the surface at night. Because of that, the whalers had no idea of their abundance or whether they might one day be a valuable resource. It was only in the 1960s that researchers from the Soviet Union began to investigate the possibility of harvesting Antarctic krill as food for animals or humans or perhaps as fertilizer. They first had to find out whether such a harvest was worthwhile and how best to do it.

It took some years of research into how to locate, catch, and then quickly process and refrigerate the creatures, which are quite delicate and quickly spoil after capture, such that they have to be processed and frozen within an hour or two of being netted. Even then, the strong and salty prawn-like taste of krill was not readily embraced by consumers, and it has taken several decades for profitable uses to be developed. Even when Japan and other nations joined in during the mid-1970s, the annual catch was only about 20,000 tons. The small size of the catch was partly due to the difficulty of processing krill in ways that would make it palatable. In the meantime, much more research was needed before scientists could estimate the total amount of krill in the Southern Ocean and the quantity that could be harvested without making the prospective krill fishery unsustainable.

How vulnerable is the krill fishery to overfishing?

Although the Southern Ocean krill fishery was first exploited commercially in the 1970s, it was slow to grow in size and value. While some of the krill catch is now consumed directly by humans, mostly as krill oil, much of the catch is processed as food for fish farms. With the increased harvesting of krill during the 1980s, there was concern that it might affect the many species that rely on it, with perhaps half of the krill biomass being eaten each year by baleen whales, seals, penguins, fish, squid, and birds. The recovery in numbers of massive baleen whales was thought to be particularly imperiled by the commercial krill fishery. This development prompted the treaty nations to agree in 1982 on the Convention on the Conservation of Antarctic Marine Living Resources (CCAMLR).

In the event, fears about overexploiting the krill fishery have not been borne out. The latest scientific estimate suggested that the combined mass of krill in the Southern Ocean amounts to 380 million tons, and the commercial fishery has never come close to netting the maximum amount that is considered sustainable by scientists. This was partly due to the collapse of the Soviet Union, which led to the withdrawal of their vessels, and to the influence of the CCAMLR in reducing the krill catch by other countries. It is also still proving difficult to develop markets for the krill.

In recent times, only about 100,000 tons of krill have been harvested annually, which is less than 2 percent of the annual catch limit recommended by CCAMLR scientists. This is positive news for the krill fishery. However, as the oceans become increasingly depleted of fish and there is a consequent boost in aquaculture to feed the world's fast-increasing population, the annual krill harvest is likely to grow. Although scientists suggest that there is scope for the annual catch to be much greater without imperiling the fishery, the impact of global warming on the fishery is more ominous and could prove impossible to control.

Why did the investigation of oil reserves take so long to initiate?

The possibility of there being oil reserves in Antarctica has long been suggested, but taking oil from Antarctica faces insurmountable problems that would make its extraction economically unviable. Until the mid-1970s, the world was faced with an oil glut as the Middle East flooded oil onto the market. The price remained so low, and new reserves of oil kept being discovered in much more accessible places, that it didn't seem worth looking for oil in Antarctica. Scientists looked anyway, as they sought to understand the nature of what lay under the ice cap and beneath the sea floor of the Southern Ocean. They were supported by governments that wanted to claim anything of value that might be found in territory they regarded as their own. In 1957, with America establishing a major base on McMurdo Sound, New Zealand undertook a geological survey of its Ross Dependency to reinforce its sovereignty and strengthen its claim to ownership of the region's resources. Companies also considered exploring the Ross Sea for oil and minerals but were dissuaded by the costs and difficulties and the uncertainty about sovereignty. Had anything valuable been found in the 1950s, it would have caused a mad rush to secure control of the deposits and scuppered moves for an Antarctic Treaty.

Instead of a mad rush for oil, there was a largely scientific investigation that steadily uncovered a picture of the continent's mountain chains, valleys, and plains and the subterranean folds beneath them that could possibly contain oil and gas. Similar seismic investigations offshore found formations beneath the sea floor that held the possibility of oil and gas being found in sizable quantities. However, it would require drilling to determine what really existed there and whether it would ever be viable to exploit these resources. Drilling would be difficult, if not impossible, on the continent with its constantly moving, crevasse-ridden ice cap and the difficulty of maintaining secure access by sea through the surrounding pack ice. That pack ice also complicated the challenge of drilling offshore, where it would be just as difficult setting up a drill rig in an ocean that froze each winter and was beset each summer with icebergs the size of small countries. It wasn't until the oil crisis of the 1970s, which saw a trebling in the price of oil and a resulting scramble to secure oil reserves in the Arctic, that governments and oil companies paid serious attention to the Antarctic region.

What has prevented the exploitation of Antarctica's offshore reserves of oil?

So long as more easily obtainable and cheaper alternatives exist, drilling for oil in Antarctica is always going to be a last resort for oil companies. The dangers and costs of exploiting oil reserves in icy seas was made clear to the world when the giant tanker *Exxon Valdez* ran aground off Alaska in 1989. Its thick oil spread across thousands of square miles of ocean and slopped ashore along 2,000 kilometers (1,250 miles) of coastline. The resulting cleanup cost the oil company billions of dollars. Had such a spill occurred off the remote Antarctic coastline, the cost of the cleanup would have been many times greater and the effects of the pollution would have been much longer-lasting in the colder seas.

The *Exxon Valdez* disaster was instrumental in convincing Antarctic Treaty nations to oppose oil and gas drilling, which might otherwise have occurred in the early twenty-first century as the world worried about dwindling oil supplies. In recent years, fresh sources of oil and gas are being exploited through fracking and other innovative means, which has allowed the United States to be self-sufficient in oil and for other nations to unlock their own reserves of fossil fuel. In the future, the Antarctic reserves will probably be saved from exploitation by the widespread adoption of electric vehicles, which will cause a steep decline in the demand for oil over the next two decades. This should lead to a slump in the price of oil, ensuring that Antarctica remains too expensive to ever be contemplated as a possible source.

How valuable is the Antarctic fishery and how can the world hope to regulate it?

It is only in recent decades that the size and value of the Antarctic fishery has come to be understood, along with its vulnerability to overfishing. The pressure on the fishery has been exacerbated by overfishing in the rest of the world's oceans, which has compelled fishing boats to head for Antarctica to harvest its previously untouched waters. They focused mainly on Patagonian toothfish and Antarctic toothfish, which are prized for their flavor and are now regulated by the CCAMLR. This required the use of fishing methods that minimize impacts on seabirds and set catch limits for each species and region. Rather than just ensuring the sustainability of the fish population, the regulations pay attention to the whole marine ecology. Although toothfish are classed as apex predators in the depths of the Southern Ocean, preying upon other fish, squid, and crustaceans, they are eaten by some whales and seals. Setting catch limits in each region must take this into account by undertaking a careful calibration of the likely impact on the complete food chain.

The scientific calculations were complicated by imperfect knowledge about toothfish. It is now known that Patagonian toothfish, which are a type of cod, live remarkable lives that can see the larger ones feeding as deep as 3,600 meters (12,000 feet) and growing to more than 100 kilograms (220 pounds) over fifty years. In addition, toothfish take about eight years to reach sexual maturity. Harvesting such a long-lived species can quickly imperil the population, as the world discovered to its cost with overexploitation of the North Atlantic cod fishery. The unregulated fishing for Patagonian toothfish in the waters off Chile and Argentina in the 1970s and 1980s saw a similar collapse of its population in some regions. Even when

CCAMLR rules were imposed after 1982, they were widely ignored by pirate fishing boats that processed their catch at sea and landed the fillets at distant ports, falsely claiming they were from other, unprotected species. It is only with the recent introduction of genetic testing and computerized tracking of legal catches, which extend all the way to the final point of sale, that illegal fishing has been markedly reduced.

Despite the regulations, catching the lucrative toothfish continues to attract pirate vessels whose owners are willing to risk tough conditions in the Southern Ocean and the chance of being detected and having their vessel seized. The difficulty of detection in such isolated waters, and the cost of pursuing any malefactors across great distances, provides pirates with a considerable degree of protection. In 2003, an Australian government fisheries ship chased an illegal Uruguayan fishing vessel for twenty-one days across the Southern Ocean, and almost to South Africa, before the vessel and its cargo of Patagonian toothfish were finally seized.

The vessel was taken to Australia, where its captain and crew were put on trial and eventually acquitted. Not so lucky was a Nigerian-flagged vessel that was owned by a Spanish company. The vessel was already wanted by authorities for poaching when it was sighted by a Sea Shepherd vessel in December 2015. It was fishing in a protected area of Antarctic waters with illegal gill nets that stretched for 100 kilometers (62 miles). The pirate vessel was chased by two Sea Shepherd ships for nearly four months across 16,000 kilometers (10,000 miles) of ocean before it was finally scuttled by its captain off the west coast of Africa, with the officers being jailed and heavily fined by authorities in nearby São Tomé and Príncipe.

Even well-drafted regulations and the declaration of prohibited fishing zones cannot prevent unscrupulous companies from taking advantage of Antarctica's isolation to poach for toothfish. So far, the treaty signatories have not been willing to devote sufficient resources to deter the poachers, while genetic and other checks on the toothfish trade have not been sufficiently rigorous to prevent illegally caught toothfish from ending up on plates in high-end restaurants. Much more effort will need to be made, using satellite technology and other means, to rid the Southern Ocean of poachers and protect its vulnerable marine life.

7

THE ANTARCTIC TREATY
OF 1959

*Why were the Russians and other claimants suddenly prepared
in the late 1950s to reach agreement on the governance
of Antarctica?*

The onset of the Cold War and the arrival of Russian whaling
fleets had raised fears that territorial rivalry in Antarctica
could erupt into armed conflict. Those fears were heightened
with the establishment of the first Russian base in the summer
of 1955–1956. Fortunately, the Russians kept their armed forces
out of the Antarctic and put whaling and science at the center
of their involvement. Much to the surprise of the Americans,
cooperation rather than competition was the Russians' watch-
word. In contrast, there was a marked lack of cooperation between the British and Argentinians. As we've seen, it got so
heated in 1952 that an Argentinian naval party fired shots over
the heads of a British expedition that was trying to reoccupy an
abandoned base on the tip of the Antarctic Peninsula. To stem
the moves by Argentina, the British government toyed with
the idea of warning them away from the Antarctic Peninsula
by threatening to test an atomic bomb there. Although cooler
heads prevailed, the race to establish bases continued. By the
summer of 1953–1954, there were eight Argentinian bases,
six British bases, and three Chilean bases established on the
Antarctic Peninsula or its nearby islands. The intensifying

competition provided more reason for the parties to choose co-operation rather than conflict.

The need for cooperation was partly driven by the nature of Antarctica. Even powerful ice-breakers could require rescue by another nation's ice-breaker, or an expedition member could need evacuation by another nation's ship or aircraft. More importantly, scientific inquiry benefited from cooperation when scientists at widely separated bases could coordinate their work and compare their observations. Such cooperation among scientists had been practiced sporadically since the late nineteenth century. It became formalized in 1958 when scientists preparing for the International Geophysical Year (IGY) established a Special Committee on Antarctic Research (later renamed the Scientific Committee on Antarctic Research), which grew to become an independent and powerful influence on the governance of the continent. This development, among others, encouraged governments to reach agreement on Antarctica's governance. Another development was the race between the Russians and Americans to be first to land on the moon. There was a fear in the 1950s that the Russians might win and claim the moon as their own, which led some Americans to suggest that a successful agreement for the governance of Antarctica could serve as a template for the sovereignty of the moon. They hoped thereby to prevent the Russians from claiming the moon by raising their flag on its surface.

There was an added impetus for such an agreement when other nations began to express Antarctic aspirations. Newly independent and nonaligned nations, such as India and Malaysia, also wanted to assert themselves in the United Nations (UN) and have Antarctica brought under their control. This push was a threat not only to the nations of Europe and the United States but also to the nations of South America as well as to Australia and New Zealand.

Matters were brought to a head by the IGY in 1958, when many countries with no previous involvement in Antarctica planned to establish scientific bases there, thus ignoring the supposed sovereignty of existing claimant nations. As a result there was a pressing necessity for a governance regime. If establishment of this regime was delayed, all those new nations with bases and activities would have to be included in the discussions. For the existing Antarctic nations, it was far better to have an agreement in place to which the newcomers would be forced to subscribe.

How did the Antarctic Treaty come about and what were its main provisions?

The establishment of permanent American and Russian bases in the 1950s helped to convince the existing claimants—Britain, France, Norway, Australia, New Zealand, Argentina, and Chile—to agree on an Antarctic Treaty while they still enjoyed some control over the matter. After nearly two years of often fractious discussion between the diplomats of the different countries, a conference was finally begun in Washington, DC, on October 15, 1959, to decide the wording of the proposed Antarctic Treaty. Apart from the seven claimant nations, the other five nations at the conference—the United States, the Soviet Union, Japan, South Africa, and Belgium—had IGY bases there. Accepting that the Russians were likely to remain permanently in Antarctica, the discussions concentrated on keeping the continent out of the control of the UN and ensuring that its governance was overseen by the nations that now had bases there. It took more than six weeks of talks, with the treaty finally signed on December 1, 1959.

Central to the treaty was a commitment to keep the continent demilitarized. This was designed to prevent armed hostilities from erupting among the increasing number of nations with bases there. Many were close to each other on the Antarctic Peninsula and its nearby islands. The Americans and New Zealanders also had adjacent bases on McMurdo Sound. Warships, air force planes, and tanks had been sent to

Antarctica in the past, and defense personnel of several nations manned their respective bases. The potential for warfare had always been present, and the arrival of the Russians increased the risk of armed conflict. Now it would be avoided.

The possibility of territorial conflict was cleverly resolved by not requiring the seven claimant nations to relinquish their historic claims and not requiring the signatories to recognize those claims. Instead, the claims would be placed "on ice" for possible deliberation at some future time. Moreover, once the treaty had been ratified, nothing that was done subsequently by any nation would be recognized as strengthening their claim. Everyone would be able to establish bases, explore territory, search for minerals, and conduct science, but only those nations that were there prior to the treaty could ever put forward a claim to ownership over any part of the continent.

The signing of the Antarctic Treaty should have brought an end to the flag raising, frenzy of exploration, and base building. However, the period between the signing of the treaty and subsequent ratification left open a window when such activities could be used to create or reinforce territorial claims. It was only after June 23, 1961, when the Antarctic Treaty finally came into force, that nothing could be done to change the status of territorial claims. At least, that was the idea. But nations have continued to embark on scientific programs, geographical exploration, naming physical features, issuing commemorative stamps, and building bases in the confident belief that at some future conference they could point to these activities as providing the basis for a territorial claim. No such conference has yet been held.

Has the Antarctic Treaty system ever been seriously challenged?

Rather remarkably, the Antarctic Treaty has not faced a serious challenge to its legitimacy over the last fifty-eight years. The stability of the treaty has been helped by having the two Cold War rivals, the United States and the Soviet Union, along with

Britain and France, among the twelve original signatories. Its stability was also ensured by the ability of new countries to accede to the treaty, provided they established a base or organized a substantial program of science. Even if they didn't make a major research effort, other nations could join as nonvoting members. This helped to appease those countries who had wanted the continent controlled by the UN and who might otherwise have described the signatories as an exclusive club intent on keeping other nations out. By the end of the 1960s, there were fifteen consultative members, which each had a vote at meetings, and one nonconsultative member whose representatives could attend meetings but not vote. At the time of writing, those numbers have since grown to include two of the treaty's harshest critics, India and Malaysia, with the treaty now having twenty-nine consultative members and twenty-four nonconsultative members.

The ban on military activities prevented Antarctica from becoming an arena for an arms race, which might otherwise have seen fortified bases being established and military maneuvers being undertaken. At the same time, the treaty does not completely ban military personnel being involved in Antarctic research and exploration or providing logistical support for bases. This helped to allay the concerns of those nations that had relied on their defense forces to supply and even man their bases. Argentina and Chile lacked a dedicated Antarctic service and used their armed forces to set up and control their bases, while the United States had used its navy and air force for logistical support and exploration.

Another provision also proved important in retaining support for the treaty. This allowed for the inspection at any time of other nations' bases. It allayed suspicions about activities being conducted at those bases, with the inspectors empowered to have access to all parts of a base to inspect equipment and ships or aircraft involved in supplying the bases. Another check against secret military activities was provided by the requirement for nations to give notice of any expeditions they were sending to Antarctica, any bases they were establishing, and any military personnel they were sending.

The truth is that the treaty would have been difficult to conclude if any valuable oil or mineral deposits had been discovered prior to it being signed. The treaty was silent about the ownership of any such resources and the manner in which they might be exploited. Neither was there any mention of the exploitation of whales, seals, and other living resources. The inclusion of clauses covering the commercial exploitation of resources would have made it more difficult to reach agreement. When these issues eventually came up for discussion in the 1980s, a proposal to allow mining and drilling for oil and gas was ruled out of bounds. Instead, a fifty-year moratorium, beginning in 1998, was imposed on such activities, with much of the world now believing that Antarctic resources should not be exploited.

Instead of the oil companies, it was the international scientific organizations and the environmental movement that have held sway over decisions relating to Antarctica. This was partly because the treaty stressed the promotion of science and encouraged cooperation rather than rivalry. Each nation agreed to disclose their planned scientific programs and the results of their experiments and observations. To further foster cooperation, scientists were to be seconded to other nations' bases and expeditions. This openness helped to strengthen adherence to the treaty and to erode some of the secrecy, suspicion, and rivalry that had characterized Antarctic activities.

Lastly, the treaty was made less objectionable by including a provision that allows its clauses to be amended. During the first thirty years of its operation, such amendment could only be done with the unanimous agreement of all the treaty signatories, which made it most unlikely to occur. After thirty years, changes to the treaty could be proposed by a simple majority of the voting representatives, which would then have to be ratified by all the consultative governments.

Although the treaty is a success, why did the proposed mining convention fail?

While environmentalists argued in the 1980s for mining and oil drilling to be banned in Antarctica, the treaty nations opted instead to create a legal framework to control the investigation and exploitation of such resources. Although the dangers and technical difficulties of operating in Antarctica ensured that any exploitation would be decades away, the nations wanted the framework in place well before any drilling or mining occurred. They already had the example of the Convention for the Conservation of Antarctic Marine Living Resources (CCAMLR), which came into force in 1982 and allowed for the sustainable harvesting of marine life. That convention provided a rough template for a mining convention, since it sought to balance the exploitation of the continent's resources with the need to protect its ecosystem, although that's more problematic when it comes to drilling for oil or mining for minerals.

The United States and New Zealand led the argument for a mining convention, with successive meetings during the 1980s thrashing out its terms. New Zealand hoped that its presumed ownership of the Ross Sea region would prove lucrative if any oil and gas deposits were developed. Its "rights" in this regard were more likely to be respected when there was still only a limited number of treaty nations. So there was jubilation in Wellington and Washington in 1988 when the nineteen consultative nations of the Antarctic Treaty signed a Convention on the Regulation of Antarctic Mineral Resource Activities. With a convention in place, companies wouldn't be allowed to embark on any ventures without paying heed to "the unique ecological, scientific and wilderness value of Antarctica and the importance of Antarctica to the global environment." The convention ensured that any activity would be done within the terms of the Antarctic Treaty system and provided for the creation of an Antarctic Mineral Resources Commission, which would have headquarters in New Zealand.

The convention seemed like a done deal until the increasingly powerful environmental movement campaigned against its ratification. The running aground of the giant oil tanker *Exxon Valdez* on the Alaskan coast in March 1989 provided a graphic illustration of the devastation that oil drilling could wreak in Antarctica. The nations that had agreed to the draft convention now began to backtrack. Australia was the first of the signatories to change its mind in May 1989, when Prime Minister Bob Hawke refused to ratify the convention and was then joined by the French government. It meant that the convention was dead in the water, since it needed unanimity to be adopted.

In place of the convention, the treaty powers met at Madrid in 1991 and drew up the Protocol on Environmental Protection, which banned any exploration or mining for fifty years from its enforcement date. The Madrid Protocol wasn't ratified by the signatories until 1998, which means it will be reviewed in 2048. It wasn't the permanent ban on mining that environmentalists wanted, but that would have been strongly opposed and perhaps even flouted by some countries. The review that's due in 2048 had held out the possibility of oil companies being able by then to reduce the difficulties of operating in Antarctica and of oil being sufficiently valuable to justify taking on those risks. While that prospect concerned environmentalists in the 1990s, there is now the realization that demand for oil will slide with a move toward electric vehicles. It can also be anticipated that in 2048 there will be even greater public support for continuing the ban.

How have nations continued to establish and reinforce their territorial claims?

The terms of the Antarctic Treaty were meant to end rivalry and reserve the continent for the pursuit of science. However, the treaty didn't prevent nations from trying to buttress their existing claims or lay the basis for new claims at a future conference on Antarctic sovereignty. The United States tried to reinforce its aspiration to control the whole continent by establishing a base at the symbolically important South Pole, by setting up a massive logistics base at McMurdo Sound, and also by undertaking the largest scientific program. The Russians responded by building several bases on the continent and on adjacent islands, including a base at the pole of inaccessibility. The symbolism of the Russian base could not compete with the symbolism of the South Pole, which had been the Holy Grail for explorers and continued to be so for modern-day adventurers. Undeterred, the Russians asserted that their territorial rights were not only based upon the original discovery of Antarctica by Bellingshausen but also upon the recent work of their ground-breaking scientists.

That sense of ownership had always been strengthened by naming its geographical features. Following the treaty, there was a renewed race to do so, as claimant nations rushed to affix their names to mountains, glaciers, and coastlines. Australia tried to outdo the Russians and Americans, who were both establishing bases in the Australian Antarctic Territory and using land-based and aerial exploration to discover and name previously unsighted parts of East Antarctica's landscape. In fact, the United States and Russia scattered their names across the entire continent, much to the consternation of countries like Argentina, Chile, and Britain that had maintained a presence on the Antarctic Peninsula and its nearby islands for decades.

It wasn't only minor geographical features that sported rival names. Nations couldn't even agree on a name for the Antarctic Peninsula, let alone its islands. Parts or all of the peninsula were known variously as Palmer Peninsula (by the Americans), Graham Land (by the British), O'Higgins Land (by the Chileans), or San Martin Land (by the Argentinians). Although the Americans and British were keen to agree on a common name for the peninsula, they had been unable to do so. It was not until 1964 that they finally agreed to call it the Antarctic Peninsula, with the northern part being called Graham Land and the southern part Palmer Land. Neutral names were also adopted for the major divisions of the continent, which were named East Antarctica and West Antarctica, while the major mountain chain connecting them was simply named the Transantarctic Mountains and the ice cap was named the Polar Plateau. However, the Chileans and Argentinians kept using the national names they'd always used.

Prior to the 1959 treaty, naming had been done according to discussions between the British Foreign Office and the U.S. State Department, with some consultation with the Australians and New Zealanders. After the treaty had been ratified, attempts were made to reach more general agreement. It took thirty years before a method was found to do so. Until then, most of the treaty nations had their own naming committees, which decided on the names that would appear on their own maps but would not necessarily be accepted by others. That began to change in 1992, when a subcommittee of the Scientific Committee on Antarctic Research (SCAR) began to compile an Antarctic gazetteer that drew upon all the competing gazetteers of various national naming authorities. To ensure its neutrality, the responsibility for compiling the *Composite Gazetteer of Antarctica* was given to a committee of Italian mapmakers, while the criteria for adopting particular names were drawn up by a committee of Germans. Even so, this effort is still a long way from compiling a map on which there is an agreed-upon name for every geographical feature.

All that the committee has been able to do is to compile a database of all the names that have been affixed to each physical feature and when they were approved by a particular nation. Twenty-two countries submitted a list of their names, together with the location and description of the features they had named. By 2007, nearly 19,000 distinct features had been included in the gazetteer. Many of these features had multiple names. The committee has been content so far to keep perfecting and updating the database as new names are submitted. Not surprisingly, at the time of writing, the existing claimant nations—Argentina, Australia, Chile, Norway, Britain, New Zealand, and France—have submitted almost half of the nearly 40,000 names. Curiously, Bulgaria had also submitted 1,335 names, while Russia had submitted 4,808 names. The greatest number by far—13,192 names—was submitted by the United States, whose history and scale of exploration reflect the extent of its territorial aspirations. If an eventual map is ever ready to be published, it is likely that American names will feature more prominently and widely than the names of any other nation. In the meantime, nations continue to publish Antarctic maps that privilege their own names.

Apart from recording names that could be used to reinforce territorial claims, nations have also recorded their own historical sites and monuments so that they can have protected status under the treaty system. Of course, the Antarctic is not very forgiving to the physical remnants of human activity, either burying them under snow and ice or floating them out to sea, but about eighty such sites have been listed for protection. Some are stone cairns, crosses, or monuments erected by the early explorers or later expeditions. Some are the wooden huts of Scott, Mawson, and other explorers that remain in various states of dilapidation. In the case of Mawson's hut, a proposal to remove it to Australia was quickly scotched in favor of keeping it intact at Commonwealth Bay so that it can provide a physical reminder of the Australian expedition and its underlying territorial claim.

While most of the monuments have been proposed and subsequently maintained by one country, some have been proposed jointly by rival Antarctic claimants, which is a credit to cooperation fostered by the Antarctic Treaty. That sense of cooperation didn't prevent Argentina proposing a flagpole at the South Pole for inclusion on the historical monuments list. The flagpole had been erected next to the American base at the South Pole in 1965 when an Argentine expedition trekked across the continent to claim the massive wedge of territory that culminates at the South Pole.

Having people buried in the soil of a territory has long been an accepted way of engendering a sense of ownership. Of course, that's difficult on the Antarctic continent, but many have died and their bodies buried there nonetheless. In the case of Scott and his companions, their bodies were left in a tent that was soon buried by snow and slowly carried by the ice toward the ocean. Others were buried beneath rocks to ensure the preservation of the gravesite as a permanent memorial. One Japanese victim even had some of his ashes returned for burial beneath a commemorative cairn at Japan's Syowa Station. Whatever the nature of the memorial or the manner of the death, these burials have helped to foster a sense of ownership over the territory in which the death occurred.

The birth of a baby in a new territory can also provide a potent symbol of ownership. That was the aim of the Argentine government when a pregnant woman was sent to one of their bases on the Antarctic Peninsula in 1978, where the first of several births was recorded on the continent. (Chile did likewise with one of its citizens on King George Island in 1984.) Argentina also made its bases more like villages with schools, churches, and shops for the families that were encouraged to live there. In 2010, there were nine families with a total of sixteen children. They made the Argentine territory akin to the Argentine mainland, which was further enhanced when a load of Argentine soil was deposited at one of the bases. Ostensibly meant for the growing of fresh vegetables, the soil had a deeper symbolic purpose of reinforcing their assertion of sovereignty.

Other assertions of sovereignty have been more commonplace. Most have concerned the establishment of permanent bases and the organization of serious scientific programs, which ensures that these nations enjoy the status of being consultative powers under the treaty. Having that status will ensure they have a say in any future division of the continent. In the meantime, there is status to be had from maintaining bases and pursuing science in Antarctica. Although scientists and politicians might argue that the science is done for its own sake, a well-developed scientific program and a large number of bases gives nations a larger voice in present-day treaty discussions and reinforces any future territorial claims.

How open is the Antarctic to tourism and adventurers?

Because Antarctica has no recognized owners, it is theoretically possible for individuals or companies to organize activities there without reference to any government. Greenpeace had a temporary base on the Ross Sea, and airline companies have flown adventurers and tourists to a tent "hotel" at the South Pole. The popularity of Antarctic tourism has increased each year, which has caused serious concern for nations with Antarctic bases, whose scientists are sometimes swamped by a sudden deluge of ship-borne tourists and face the ever-present risk of being asked to rescue tourists or adventurers from the terrors of the environment. It's an unwanted imposition and an onerous burden on their hard-pressed Antarctic budgets.

Although the first cruise ship in 1966 had only fifty-eight American tourists aboard, some three thousand tourists visited in the summer of 1987. As even bigger cruise ships began heading for Antarctic waters, the possibility of a human and environmental catastrophe increased. The *Lindblad Explorer*, which had been built in 1969 specifically for polar tourism, ran aground in Antarctica in 1972 and again in 1979, with the passengers being evacuated and the ship successfully refloated. Not so lucky was the Argentine supply ship *Bahia Paraiso*, which gashed its hull on a reef in 1989 as it was steaming toward America's Palmer base. The 234 passengers and crew were rescued and accommodated at the base, while 600,000 liters of diesel fuel polluted the icy waters, affecting the local penguin rookery and some bird species.

In December 2007, the *Lindblad Explorer*, since renamed the *Explorer*, came to grief for the third time as it was steaming through the Bransfield Strait off the Antarctic Peninsula. The captain was unfamiliar with Antarctic conditions, which feature thicker and harder ice than in the Arctic, and he went too fast into ice that tore a gash in the vessel's side. Fortunately, the sea was calm and rescue ships were able to reach the lifeboats within five hours of the evacuation. It was fortunate, too, that the ship sank in very deep waters, which helped to limit the harmful effects to wildlife. Not so lucky were the passengers and crew aboard the Air New Zealand sightseeing flight in 1979 that crashed into Mt Erebus, killing all 257 passengers and crew.

The Antarctic Treaty powers tried to bring some controls to the burgeoning industry. While most of the isolated bases in East Antarctica were not visited, the scientists at America's South Pole station often found their work interrupted in summer by the arrival of adventurers trekking in the footsteps of Scott and Amundsen or by small groups flying in to be photographed at the South Pole. Scientists on the Antarctic Peninsula or its nearby islands could find themselves with hundreds of tourists landing ashore, crowding around their buildings, and roaming across the landscape photographing the wildlife.

A basic code of conduct introduced for tourism operators in 1966 was made progressively more restrictive over the following three decades as tourism numbers rose and the instances of misadventure and damage increased. Historical sites were looted by souvenir hunters, and some fragile ecosystems were damaged. When scientists had to help with the rescue and care of tourists whose ships had come to grief, it could destroy a year's carefully planned scientific program. As a result, from 1994 organizers were required to give notice of their visits and to obey the directions of the station commanders. They were also limited to where they could visit, so that more pristine areas could be protected from the impacts of mass human contact. By the beginning of the twenty-first century, as giant cruise ships with thousands of passengers sailed within sight of the continent and an armada of smaller expedition ships disgorged tourists ashore, the guidelines were tightened even further. All tourist companies were required to have contingency plans and insurance in place to cover the possibility of search and rescue.

The restrictions were supported by the tourist companies, since it was the pristine nature of Antarctica that provided the major appeal for their customers. The International Association of Antarctic Tourism Operators, which was formed in 1991, urged its members to abide by the guidelines. More effective moves were made by the International Maritime Organization, which banned cruise ships from carrying heavy oils in Antarctic waters because of the greater environmental damage the oil could cause. Insurance companies also refused to cover any of the larger ships with unstrengthened hulls that wanted to venture into Antarctic waters, while ships with more than five hundred passengers were not allowed to land tourists ashore. Nevertheless, each summer still sees more than 40,000 tourists visit Antarctica.

Adventurers have proved harder to regulate. Some are intent on climbing the highest mountains on every continent, while others want to emulate the feats of early explorers by trekking to the South Pole or sailing like Shackleton in an open boat from Elephant Island to South Georgia. There has even been a marathon organized in Antarctica, albeit on King George Island. There have been so many adventurers and South Pole sightseers that a private airline was established to cater to their needs. The Canadian-owned Antarctic Airways first flew an aircraft to the continent from Argentina in 1987. Using an old propeller-driven DC-4, the plane took passengers to the continent's tallest peak, Mount Vinson, where a base was established on the ice cap, while smaller aircraft took sightseers for brief visits to the South Pole. Hundreds of mountaineers and adventurers used the airline, while it was also used by scientists going to bases and by smaller nations with limited logistical support to take supplies and personnel.

How have nonstate actors used the Antarctic Treaty to assert their own claims?

Once the Antarctic Treaty was ratified in 1961, other groups began to seek a voice in the governance of the continent. Environmental groups wanted to protect the continent from the damaging activities of oil companies and from those nations rushing to establish bases and undertake scientific programs without regard to the environmental impact. At some bases, rubbish was dumped into the sea, while the United States even operated a small atomic power station at its McMurdo Station. In 1978, more than two hundred environmental groups joined together to form the Antarctic and Southern Ocean Coalition, which became an influential lobbying group based in Washington and was accorded observer status at treaty meetings.

In 1987, the international environmental group Greenpeace set up a temporary base on Ross Island, close by America's huge base at McMurdo Sound and New Zealand's more modest affair. The base was set up in the context of moves by mining and oil companies to explore the resource potential of Antarctica, with the presence of Greenpeace activists designed to guard against such activities and signal to the treaty powers that they could not ignore the rising tide of environmental concern. They began by inspecting the adjacent American and New Zealand bases, where sewage and rubbish was tipped into the sea and where the Americans had buried radioactive waste in the ice. Greenpeace inspected as many bases as they could and then wrote a damning report describing what they had seen. It helped to shame some, but not all, of the bases to clean up their surroundings and return rubbish to their home countries.

When the French decided to build a landing strip on an island off Adélie Land in 1989, the Greenpeace activists sailed to the rescue of the penguin colony that was being dislodged by the French bulldozers. Putting themselves in the path of the bulldozers, the activists were successful in stopping the airstrip from being completed and revealing the idiocy of building a runway through a penguin rookery that was being studied by scientists. The Greenpeace action reminded the treaty powers of their obligation to protect Antarctic wildlife, and that responsibility became an increasing part of the regular treaty discussions.

The harvesting of krill and fish raised fears that some species of fish were being driven to extinction. To control the fishing, a conference in 1980 led to the Convention for the Conservation of Antarctic Marine Living Resources (CCAMLR) and a scientific organization to oversee it. The CCAMLR included nations that were not signatories to the treaty but were engaged in fishing and the harvesting of krill. It was an attempt to balance the desire of some nations to exploit Antarctic resources with the desire of others to preserve the region. Of course, the approval of any fishing causes inevitable damage to the environment, but the CCAMLR has allowed for a tightening of restrictions and even the creation of marine parks.

The tightening of rules on tourism and the exploitation of marine resources was done amid calls by the environmental movement for Antarctica to become a world park and by scientists for it to be reserved for science. Although the first members of SCAR were nominated by their respective countries, most of them quickly proved their independence and asserted their own views as to what activities should be permissible and how scientific programs should be coordinated. Recently, SCAR's remit has been broadened to include all the Southern Ocean encompassed by the Antarctic Circumpolar Current and even those sub-Antarctic islands that lie north of that current's reach.

How successful has the Antarctic Treaty been?

Prior to 1959, the seven nations with territorial claims—Britain, France, Norway, New Zealand, Australia, Argentina, and Chile—couldn't even agree among themselves about their respective claims, let alone have them recognized by the international community. By setting those claims to one side, the treaty allowed all nations with a substantial interest in the Antarctic to reach agreement on many other Antarctic issues. Consensus replaced the confrontation that had marked the 1940s and 1950s, and other nations that wanted to become involved in Antarctica could do so as part of the all-embracing treaty system.

By taking sovereignty off the table, the treaty nations opened the possibility of a nontreaty nation or a nongovernmental body defying the decisions of the treaty powers. But that hasn't happened. Because the signatories to the treaty include almost all the world's major and middle powers, it would be foolhardy for any nation, company, or group to act in defiance of the treaty and its protocols. Even the private airline that takes adventurers to Antarctica does so with the tacit support of the treaty powers, since it couldn't operate without using airports in Chile or Argentina. Treaty powers are prepared to tolerate its activities, because the planes also transport scientists and supplies to some of their Antarctic bases.

The relative success of the treaty over more than half a century is remarkable, but it is made more powerful for having the support of the international scientific community and influential environmental groups. The greatest threat to the observance of the treaty has been from the signatories themselves, who have been lax in following its terms, particularly in relation to protection of the environment. Some nations are still refusing to comply with the environmental clauses. Their bases are littered with abandoned machinery and equipment and their waste has been allowed to flow into the sea. One Russian base has been abandoned altogether, leaving barrels of fuel to rust away.

Does anyone still see a strategic value in trying to exercise
sovereignty in the Antarctic?

Today, there is some strategic value to be had from controlling the maritime gateway between the Atlantic and Pacific
Oceans. Yet the Antarctic Treaty restricts the use of scientific
bases for military purposes. That doesn't mean they can't be
used to maintain surveillance on the activities of other nations.
With the rising temperatures that West Antarctica has experienced over the last decade, the region has become that much
more habitable and valuable as a possession if its sovereignty
ever again comes up for discussion. By having bases there, nations increase their right to claim ownership of an island or islands or part of the peninsula, along with its territorial waters
and whatever lies beneath them.

The strategic value of East Antarctica might not seem as
obvious, but that hasn't prevented nations from maintaining
permanent bases. As in West Antarctica, the potential wealth
beneath Antarctic waters is one reason why nations have
been prepared to bear the considerable expense of having an
Antarctic presence. There's also the prestige that accrues to nations in Antarctica, which is a factor that influenced the dispatch of expeditions from the late nineteenth century onward.
For Australia, the strategic value of the Australian Antarctic
Territory, comprising 42 percent of the continent, is driven
largely by a ridiculous fear that it could fall into the hands of
a potential enemy, whether it be Japan, Russia, or China, and
thereby somehow imperil the southern approaches to Australia
despite it being more than 4,500 miles (7,200 kilometers) from
Antarctica.

Antarctica has also become more desirable because of global
warming. Not only does its ice hold the history of previous
warming events, and therefore has value for predicting the
Earth's climate future, but the prospect of there being an ice-
free continent in the distant future makes it desirable to own
Antarctica. Whereas there was talk in the 1950s of detonating

nuclear bombs in Antarctica to melt the ice to make the continent habitable, there is now the possibility of global warming causing the ice to melt over the next few centuries. An ice-free continent would be a valuable acquisition, particularly for nations whose home territory had become less habitable because of that same global warming. While the Antarctic Treaty has largely kept these acquisitive urges under wraps, they are nonetheless real and help to motivate the governments that continue to commit resources to Antarctic research.

8

GLOBAL WARMING

How vulnerable to climate change are Antarctica's fauna?

Antarctica's fauna are very vulnerable to climate change. All we have to do is look at the penguins. With the recent dramatic loss of ice in the Arctic, the world has focused on whether polar bears are facing extinction. Less attention has been paid to penguins in Antarctica, which face challenges of their own. The species of penguin that live south of the cold current that circles Antarctica are the Adélie, king, emperor, chinstrap, southern rockhopper, eastern rockhopper, macaroni, gentoo, Ellsworth's gentoo, erect-crested, and royal penguins. Most live on the islands rather than on the continent. Emperor penguins are the largest species, and they live on the ice attached to the continent and its islands, with their forty or so breeding colonies sometimes located up to 100 kilometers (60 miles) from open water. In the past, there was always the danger of the sea ice extending further than normal, which could leave a colony of penguins too far from the water to feed. In such circumstances, a whole colony could perish.

With global warming, a new danger has emerged. Massive glaciers or vast stretches of ice shelves are breaking away because of the warming water and drifting into positions between colonies and the water. This was the fate of a colony of Adélie penguins, which weigh just 5 kilograms (11 pounds) and nest on sites that are free of snow and ice. In 2010, a glacial tongue that was almost 3,000 square kilometers (1,200 square miles) in size broke off and floated into Commonwealth Bay on the Antarctic coast south of Australia. Sea ice froze between the iceberg and the shore and did not melt in summer. This had a dire effect on the colony of Adélie penguins, numbering around 160,000, which now had to trek 60 kilometers (37 miles) to reach the sea to feed. Since then, the number of penguins in the colony has been reduced to just 10,000. The colony could die out completely if the iceberg remains in place and makes the usual breeding ground effectively uninhabitable.

Of course, global warming is more prone to reduce the extent of sea ice rather than increase it, which brings a threat of a different kind. Because almost all emperor penguins live, breed, and molt on so-called fast ice that is connected to the continent, they rely on that ice being present for about eight months of the year. If higher temperatures cause the ice to be slow to form in the winter, or to break up earlier in the summer, the penguins' breeding cycle is imperiled. If the ice breaks up before their molting has concluded, and they're forced into the water, they will not survive. This seems to have been what happened to a small colony of emperor penguins that were recorded in 1948 on the sea ice of the appropriately named Emperor Island, located in Marguerite Bay off the coast of the Antarctic Peninsula. The colony had a stable population of about 150 breeding pairs between 1948 and 1970, after which the colony went into a slow decline until no penguins were seen there by 2009. These latter decades were a time when the Antarctic Peninsula experienced a 2°C increase in temperature, which shortened the period during which sea ice was present.

While the emperor penguins live much of their lives on the upper surface of the sea ice, the underside of the ice is the habitat for microorganisms and juvenile krill, which provides an important food for the emperor penguins and for the fish and squid on which they also feed. As a result, the higher temperatures have a double impact on the viability of emperor penguin colonies, since the reduced sea ice results in less krill as well as less fish and squid. There may also be a third impact, with scientists suggesting that the warmer temperatures may have allowed predatory seabirds to breed earlier and then prey upon the emperor chicks when they're at a more vulnerable size.

Adélie penguins are particularly reliant on krill for food. Less sea ice in the winter means less krill in the summer. A concomitant increase in snowfall results in a spring melt that floods the penguin nests. Since the 1980s, scientists have noticed a precipitous fall, about 70 percent, in the Adélie penguin population on Anvers Island, amounting to the loss of some ten thousand breeding pairs. Meanwhile, chinstrap and gentoo penguins have taken advantage of the warmer temperatures to extend their range from the islands off the Antarctic Peninsula and onto the peninsula itself. Because they establish their breeding colonies on exposed land, the chinstrap and gentoo penguins are not dependent on sea ice to sustain their life cycles. They also have diets that are less dependent on krill, so if a reduction in sea ice causes krill to be scarce they simply increase the proportion of fish and squid in their diet. At least in the short term, wider dietary options have allowed these penguin species to thrive. King penguins have also increased in numbers on such sub-Antarctic islands as Macquarie Island, where they were harvested almost to extinction in the early twentieth century and now number more than half a million. For reasons that are not clear, rockhopper penguins living alongside them have not been so fortunate and have declined in number.

These are just some of the effects that global warming has had on penguins over the past fifty years. Scientists have had difficulty coming to definitive conclusions about their fate because there are limited places that allow scientists to closely observe and count penguins over an extended period of years. The easiest places to do so are near scientific bases, where penguin colonies are often located due to the suitability of the snow-free terrain. Yet despite their proximity, calculating penguin populations has been fraught with difficulty. The sex of some species is difficult to determine by sight, and the practice of counting breeding pairs has caused scientists to seriously underestimate their numbers. Because nonbreeding penguins are usually out at sea, scientists have omitted them from their estimates of the total penguin population.

It was only when Australian scientists used aerial and ground surveys, along with time-lapse photography, that they realized their long-held estimate of the total population of Adélie penguins in East Antarctica was less than half the new estimate of about 6 million penguins. While that certainly is a substantial number, the Adélie population could easily plummet as the changing sea ice conditions in East Antarctica gradually come to replicate the conditions in West Antarctica. Much more work needs to be done to understand the whole Antarctic ecosystem, how it is responding to the pace of global warming, and the effects on various penguin species living in very different habitats.

How fast is the ice cap melting and what effect is it having on the sea level?

It depends on your definition of "fast," but ice cap melting is certainly causing sea levels to rise. Over the last century, sea levels have risen by 10–20 centimeters (4–8 inches). Although the rate of increase is accelerating, it still amounts to an almost imperceptible rise of 3.2 millimeters (0.13 inches) per year. Apart from some low-lying islands that have experienced inundation, the effects have been barely noticeable in most parts of the world. However, that is set to change as global warming increases and the sea level rises even more precipitously. The latest predictions suggest that the rise by the end of this century will be somewhere between 30 centimeters and 2.5 meters (1–8 feet). The wide range in the computer modeling reflects the current uncertainty as to whether or not the world can reduce its output of greenhouse gases.

Global warming affects the sea level in two ways. One is by the warming of the oceans and the consequent swelling of their volume. But the greater increase to the world's oceans will come from the release of the water that is currently stored in the glaciers and ice sheets of Greenland and Antarctica. If these were all to melt, it would inundate many of the world's biggest cities and flood some of the most productive farming lands. The increased temperatures from unrestrained global warming would also make much of the world uninhabitable to humans. Although the complete melting of the Antarctic ice cap could take several thousand years, increasingly harmful effects from the present rate of global warming are going to be felt over the coming decades.

As the world has become hotter over the last century, there has been a noticeable diminution of glaciers, from Europe to South America, and a dramatic reduction of winter sea ice in the Arctic. More worrying has been the evidence from Greenland, where there have been signs of increased melting of its ice sheet and faster movement of some of its glaciers into the sea. Although Greenland has lost much of its ice sheet during previous warming events more than a hundred thousand years ago, it has never experienced the present rate of warming. Scientists are still unsure about the implications of warming, with some suggesting that the collapse of the ice sheet could come much faster than previously predicted. If the ice sheet did melt completely, sea levels would rise by more than 7 meters (23 feet). Long before that happened, the increased icy water from Greenland would have the potential to reverse the flow of the Gulf Stream that warms Britain and northern Europe, with obviously calamitous effects on those places.

While the melting of the Greenland ice sheet would have disastrous effects on the world, a much greater contribution to sea level rises would come from a melting of the Antarctic ice sheet, because it contains 90 percent of the world's fresh water. Although it would take many centuries to melt completely, scientists have estimated that the melting could contribute an additional rise of 15 meters (49 feet) by 2500. Left unchecked, the melting would eventually go on to cause a rise of about 65 meters (213 feet). Even a partial melting would have a serious effect on the sea level, and there are worrying signs that it is starting to happen, with the Antarctic ice sheet being responsible for about 10 percent of the recent sea level rise. Without a drastic reduction in the production of greenhouse gases, it will get much worse.

Researchers still have much to learn about how the warming of the world is affecting Antarctic weather systems, ocean currents, ice formation, and glacial flows. Each winter, the seas around Antarctica freeze over and then melt again in the summer. The extent of the winter sea ice varies each year. In September 2014, it reached 20 million square kilometers, which was about 50 percent bigger than the area of the continent and represented its highest extent since satellites first began photographing Antarctica in 1979.

The more sea ice there is, the greater its ability to reflect the sun's rays. Rather than allowing heat to be absorbed by the ocean, expanded sea ice could moderate the rate of global warming. In great contrast to the situation in the Arctic, this is what seemed to be happening in Antarctica in 2014, with the expanded sea ice in the Southern Ocean partially offsetting the warming effect of the Arctic ice loss. However, that completely reversed in the summer of 2016–2017, when the sea ice melted to its lowest extent since 1979, covering little more than 2 million square kilometers (770,000 square miles), which meant that the Southern Ocean was absorbing more of the sun's summer heat. This causes an even greater melting of the sea ice and the undermining of the ice shelves attached to the continent. It won't be known whether or not this constitutes a new trend until future satellite observations have been made.

Until recently, scientists were puzzled by the temperature differences between East and West Antarctica. In West Antarctica, comprising about a third of the continent and including the Antarctic Peninsula, temperatures have risen by about 2°C over the last half-century, making it one of the fastest-warming regions of the world. It's not only the sea ice of West Antarctica that has been reduced by this warming but also the ice sheet on land, which has shown signs of increased melting into the surrounding ocean. In contrast, East Antarctica seemed to be getting colder and experiencing heavier snowfalls, which allowed scientists to hope that the ice melt in West Antarctica might be offset by the apparent addition of snow to the ice sheet of East Antarctica. If they were right, it might have meant that Antarctica was not contributing much to the observed rise in sea levels and might even have been partially offsetting the sea level rise being caused by warming in the Arctic.

Unfortunately, recent observations have suggested instead a much more worrisome scenario. Rather than the circumpolar current insulating East Antarctica from the warming ocean further north, observations from deep below the ocean surface have revealed that relatively warm waters have been eating away at the base of the Totten Glacier, which draws ice from an area larger than California. The observations by Tasmanian researcher Stephen Rintoul on board the *Aurora Australis* confirmed earlier satellite observations about the thinning of the glacier and its floating ice sheet, which is losing about 10 meters of its thickness each year. At nearly 2.4 kilometers (1.5 miles) thick, the glacier has got a lot to lose, and its slow melting will not, by itself, lead to a measurable change in the sea level. After all, the melting is occurring at a point where the ice is floating and cannot therefore affect the sea level. However, the glacier and its ice sheet perform a vital function in holding back the ice cap behind it, which is slowly sliding toward the sea. As the bulk of the glacier and its floating ice sheet is thinned by melting, its ability to hold back the vastly greater ice sheet will be gradually reduced. This will occur over a centuries-long process that has the potential to cause an estimated rise in the sea level of 3.5 meters (11.5 feet), which would flood much of Manhattan. And that's just from this part of the ice sheet.

Why is global warming affecting parts of the continent in dramatically different ways?

In this, Antarctica is no different from the rest of the world. The variation has been seen very dramatically on the Antarctic Peninsula, where one weather station observed a steady rise in temperature of nearly 3°C during the second half of the twentieth century, much greater than anywhere else in the world, only to have the temperature reverse and begin to cool at a similar rate. Scientists were puzzled by the extent of the rise and the reason for the sudden reversal. It seems that there were several causes. One reason for the warming was the onset of westerly winds caused by El Niño conditions in the Pacific, which brought warmer air and pushed the sea ice from the peninsula, rather the usual easterly winds that bring colder air and prevent the dispersal of sea ice. Another reason was the thinning of the ozone layer in the latter part of the twentieth century, which also encouraged the warmer westerly winds. Once international action caused the ozone layer to thicken, and La Niña conditions were dominant in the Pacific, there was more encouragement for the resumption of the colder easterly winds.

The warming of the region during this period had spectacular effects, causing the shrinking of glaciers and ice shelves, the reduction of sea ice, the encouragement of plant life, and the collapse of much of the massive Larsen B ice shelf in 2002. Located on the Weddell Sea, the Larsen Ice Shelf is attached to bays along the peninsula. The shelf is divided into seven distinct parts. Larsen A was further north before disintegrating in 1995, while Larsen B is about the size of the U.S. state of Rhode Island. For at least ten thousand years, Larsen B had been replenished from the continent's ice sheet, only to collapse and disintegrate within the space of three weeks. Its rapid disintegration shocked scientists and was a direct result of the rise in the temperature of West Antarctica during the previous half-century. The implications of its disintegration are momentous and ominous, since as noted above the shelf had acted as a brake on the movement of the ice sheet toward the sea. Now that the shelf is gone, scientists have observed one of its glaciers accelerating until it has begun moving at three times its former speed. This is what glaciologists and climate scientists long feared.

Further south, the much larger Larsen C ice shelf has been showing signs of going the same way. At 50,000 square kilometers (20,000 square miles), the Larsen C is Antarctica's fourth-largest ice shelf, and it now has a deep rift more than 100 kilometers (62 miles) long, which is causing about 10 percent of the shelf to break off. Once completed, the calved ice will form one of the world's largest-ever icebergs. At 5,000 square kilometers (1,900 square miles), it will be bigger than the U.S. state of Delaware. Further disintegration of Larsen C will ensue, as the warming ocean water eats away at the floating bottom and increasing meltwater causes further rifts to form and extend their way across the remaining shelf. Once this restraining influence has been removed, the slowly shifting ice sheet behind it will likewise accelerate movement toward the Weddell Sea and cause the oceans to rise even higher. The acceleration is also assisted by increasing streams of meltwater that flow into crevasses and find their way to the base of the ice sheet, where the water acts as a lubricant, allowing the ice above it to slide significantly faster.

On the west side of the Antarctic Peninsula, scientists have been using satellite data to understand the behavior of five massive glaciers that flow into the Amundsen Sea. Although they have discovered great disparities between the different glaciers, all of them have shown a tendency to accelerate as warm water eats away at the ice that connects the underside of the glacier to the land. This has caused a marked thinning of the glaciers, with one of them losing up to 7 meters (23 feet) a year in thickness. More worrisome, this thinning has worked its way up the glaciers as the accelerating movement causes more ice to be taken from their hinterland. In the case of one glacier, the thinning has spread for hundreds of kilometers, which means that it is pouring an ever-increasing amount of ice into the ocean. Combined, it is estimated that these five glaciers alone are putting up to 140 billion tons of ice into the Southern Ocean each year, which is raising the world's sea level by about 0.4 millimeters (0.016 inches) per year.

The Ross Ice Shelf is the largest of all the Antarctic ice shelves, being nearly the size of France. It is where Robert Scott's party died and from where Roald Amundsen launched his successful bid for the South Pole. Being much further south than any of the others, and therefore much colder, the Ross Ice Shelf is the most resistant to global warming and should be able to continue slowing the movement of the several glaciers that feed into it. Although it is still too cold by a few degrees for serious disintegration to occur, it will disintegrate and release a surge of ice once the air and water of the Ross Sea becomes sufficiently warm. Its disintegration won't happen anytime soon, but scientists have discovered that disintegration events over the last several million years occurred very suddenly and caused a rapid rise in sea level. The eventual collapse of the Ross Ice Shelf will be preceded by the collapse of the Filchner-Ronne Ice Shelf on the opposite side of Antarctica. Encompassing the southern coast of the Weddell Sea, it is almost the same size as the Ross Ice Shelf and is predicted to disappear by the end of this century.

As the rapid breakup of Larsen B and the retreat of the glaciers on the Amundsen Sea have revealed, the melting in West Antarctica poses the most immediate danger to the world. Until recently, there was some hope that a partial reversal in temperature in West Antarctica might stop or even reverse some of these changes, perhaps causing the Larsen B to reform as the cold easterly wind system returned. However, the reappearance of Larsen B is unlikely to occur, and even if it does it will be a brief hiatus in the inexorable warming of the region and of the planet. The hitherto remorseless buildup of greenhouse gases around the world will continue to wreak changes to the temperatures of the oceans and to the complex weather systems and ocean currents that determine climate outcomes in Antarctica.

The experience of the Larsen ice shelves in West Antarctica has serious implications for the ice shelves of East Antarctica and the enormous ice sheets that those shelves help to hold back. The winds and ocean currents seemed to be quarantining East Antarctica from the worst effects of global warming, with the coldest temperature ever recorded (–89.2°C at Russia's Vostok Station in 1983) being exceeded when satellite data revealed that a new low temperature of –94.7°C had been reached in 2010. That gave scientists some hope that the polar ice cap might not experience the melting and disintegration seen in Greenland and that it might be replenished with snow faster than it was losing mass by slipping into the sea. Thus the cold circumpolar current would keep the warming waters away from the ice shelves of East Antarctica. If that hypothesis was borne out, it could have meant that East Antarctica would not contribute to the rising sea level, as moisture from the ocean falling as snow would be greater than the ice melting into the ocean. However, those hopes have been dashed after satellite observations confirmed that the total mass of all the Antarctic ice sheets is declining, with even those in East Antarctica showing signs of thinning. No matter how much snow is being added, it will not be sufficient to balance the ice that is being lost.

How is climate change affecting Antarctica's ecology?

Global warming has caused complex changes to many Antarctic ecosystems, with some life forms struggling to cope with the nature and rapidity of the changes. Dramatic changes to the extent of sea ice can have serious effects all the way along the food chain. The sea ice covers about 20 million square kilometers (7.7 million square miles) for much of the year and shelters the algae and other microscopic life forms on which krill depend for food. If there is a drastic reduction in sea ice, there is less food for the krill and therefore less krill for the larger life forms that feed upon it, which range from fish and penguins to seals and whales. Life becomes especially difficult for those animals that are especially dependent upon krill, such as Adélie and emperor penguins, although a couple of species of penguins have increased their numbers and extended their habitats in response to the warming of West Antarctica.

While some scientists have concentrated on changes caused by global warming to the larger life forms, it is the changes to the most minute life forms that will likely have the most significant effect on the biology of the Southern Ocean and could even feed back to affect the pace and direction of global warming. Some of the smaller life forms, called phytoplankton, are smaller than a pinhead in size, yet these single-cell marine plants play a crucial role at the base of the food chain. They can also play a role in moderating the pace of global warming by removing carbon dioxide from the atmosphere and producing chemicals that encourage the formation of clouds. Those vital functions could be imperiled by higher water temperatures, less sea ice, and increased acidification of the ocean. It is too early to know how this will play out in practice, as some types of phytoplankton might do better than others. The larger ones favored by krill are expected to suffer, which will adversely affect all the life forms that rely upon krill, while another smaller species is predicted to thrive and, because of its role in the formation of clouds, may thereby help to moderate the warming. The outcome is likely to depend upon the rapidity of the changes in the Southern Ocean and the ability of these microorganisms to adapt to the more challenging conditions.

How will the changes in the Southern Ocean affect
the Antarctic ecosystem?

As global warming increases the concentration of carbon dioxide in the atmosphere, scientists have discovered that it is causing the world's oceans to become more acidic. This change appears to be happening at the fastest rate in 300 million years. About a third of the carbon dioxide produced by the burning of fossil fuels is being absorbed by seawater and converted into carbonic acid, resulting in the increased acidity of the oceans. Over the past two hundred years of industrialization, the oceans have experienced an increase of about 25 percent in their acidity. If this problem is not stemmed, it is likely to have profound consequences in the Southern Ocean, since cold, storm-tossed waters absorb more carbon dioxide than warmer water, and the chemistry of colder water causes the rate of acidification to be 50 percent faster than in the tropics. It has been predicted that some of the fundamental life forms of the Antarctic region, such as krill, will be hit hard by the worsening acidification and that this will then affect all the larger life forms that feed on krill. Along with a range of seabirds, baleen whales that specialize in feeding on krill, including blue whales and minke whales, will have their survival threatened.

Scientists are still studying how krill might be affected. The greater acidity of the surface water makes them vulnerable since that is where their life cycle begins with the spawning of their eggs, which then drop down to hatch at depths ranging up to 1,000 meters (3,280 feet), and from there they swim back to the surface to begin feeding on algae. Laboratory testing in water tanks, with various levels of carbon dioxide pumped through them, has found that the eggs don't hatch in water that has a level of 2,000 parts per million. The present level of carbon dioxide is well short of that at 380 parts per million (ppm). However, if nothing is done to stem the rise in carbon dioxide levels, it is predicted to reach 1,400 ppm by 2100, by which time as much as 50 percent of the Antarctic krill will be unable to hatch. At the same time, the increased acidification of the ocean will be impinging on other stages of the krill's life cycle, making their shells thinner and softer. By 2300, if carbon dioxide continues to increase at the present rate, and the krill prove unable to adapt to the changing conditions, they will disappear from the Southern Ocean. Their absence will signal the end of the rich Antarctic ecosystem. It will also have direct consequences for human beings, since krill have been incorporated into our own food chain over the last half-century, following their harvesting as food for farmed fish.

9

THE FUTURE OF ANTARCTICA

What agreements are in place for the protection of Antarctic marine life?

There are three agreements that are designed to protect marine life in the seas around Antarctica. The first one to be concluded was the Convention for the Conservation of Antarctic Seals (CCAS), which was signed in 1972 by seventeen signatories to the 1959 Antarctic Treaty. Rather than declaring an outright ban on the killing of seals, the CCAS provided for their killing in a sustainable manner and without harming the overall ecology of the Antarctic region. In fact, no exploitation of the seal population occurred under the convention, other than the sampling of small numbers of seals by scientists for research purposes. Other forms of marine life are protected by the second agreement, the Convention for the Conservation of Antarctic Marine Living Resources (CCAMLR), which was drawn up at a conference in Canberra in 1980. Twenty-four nations and the European Union are now parties to the CCAMLR. Like the CCAS, it was not designed to ban the harvesting of marine life but rather to provide a legal framework that would ensure these practices were done according to the best scientific advice.

The CCAMLR was agreed after concern about the possible overexploitation of krill. Similar concerns were expressed about the possible overexploitation of fish and squid. The CCAMLR established a commission based in Hobart that would set limits to any harvesting, advise on its likely impacts, and initiate an inspection regime to oversee fishing operations in Antarctic waters. The CCAMLR has also created marine protected areas (MPAs) in the Southern Ocean in which all fishing is banned. The first such MPA was declared in 2009 and covered 94,000 square kilometers (36,000 square miles) off the South Orkney Islands, while a second protected area was declared in 2016 and covered 1,550,000 square kilometers (600,000 square miles) of the resource-rich Ross Sea. Most of the area is to be completely free of fishing, even for scientific purposes, while 28 percent of the area is available only for scientific research. The commission also introduced measures to minimize the killing of seabirds as an accidental byproduct of fishing operations and ensure that fishing boats used methods that minimized the amount of unwanted by-catch that was caught in nets, particularly through the bottom trawling of the seabed. Because of the research undertaken by the CCAMLR scientific committee, limits were set for the annual amount of krill that could be harvested in various parts of the Southern Ocean. Total bans were imposed on the catching of Patagonian and Antarctic toothfish, which are vulnerable to overfishing because of the long time it takes for them to reach maturity.

The toothfish have come under intense pressure from commercial fishing, since they are highly prized by upmarket restaurants and are slow to reproduce. Although CCAMLR inspectors were given authority to board suspicious fishing boats, they had to find them and rely upon the captains of those fishing boats to respect their authority. There was usually too much at stake for the pirates to do so. Although pirate boats have been sunk or seized, it was estimated at the time of writing that there are still at least seven pirate boats operating in contravention of the convention. The commission has responded with measures to supervise the trade in complying fish from the point of catching to the point of final sale, which then allows noncomplying fish to be identified and seized. Yet the illegal fishing goes on.

In regard to whales, the treaty powers left the regulation of whaling in the Southern Ocean to the existing International Whaling Commission (IWC), which was established in 1946 and composed of the fifteen whaling nations. There was little concern for whales, since commercial whaling stopped altogether in 1982 after the IWC imposed a worldwide moratorium and in 1994 declared a whale sanctuary of some 31 million square kilometers (120 million square miles) in the Southern Ocean. Only Japan continued to hunt for whales in Antarctic waters, ostensibly for research purposes, with the numbers killed being relatively small and restricted to the numerous minke whales. Japan's scientific program was challenged by Australia in the International Court of Justice in 2014. When the decision went against Japan, it refused to abide by it and announced four years later its intention to leave the commission altogether and resume commercial whaling, which is likely to take place in the north Pacific rather than the Southern Ocean.

The prolonged disputation over Japanese whaling revealed the IWC's lack of power to enforce its regulations when one or more countries are determined to defy its decisions. Unlike the consultative countries to the Antarctic Treaty, the IWC is not based on an international treaty with the legal force that would entail for its signatories. Members of the IWC don't even have to have any involvement in whaling. In fact, most of them haven't been involved and have no intention of becoming involved. That makes the IWC very different from the Antarctic Treaty, whose signatories must demonstrate a serious and practical commitment to scientific research in Antarctica. It means that the protection of whales in the Southern Ocean is based largely on political considerations. Although Japan's small whaling fleet is unlikely to return to the Southern Ocean once it resumes commercial whaling closer to home, there is always the risk of another country deciding to do so. The best hope for the whales could be for the IWC to cede control over whaling in the Southern Ocean to the CCAMLR commission so that decisions about whaling were based upon scientific considerations, which might see scientists decide that the abundant minke whales were able to be harvested sustainably.

Does mass tourism pose a threat to Antarctica?

There are no immigration officers on hand to greet visitors to Antarctica. Nor are there any police to enforce the regulations designed to protect the environment and prevent interference with scientific research. Yet more than a million people have stepped ashore in the nearly two centuries since the American sealer John Davis became the first person to land on the Antarctic Peninsula in 1821. Each summer, some 40,000 tourists visit Antarctica and its adjacent islands, greatly outnumbering the annual influx of scientists. When a crowd of tourists walk thoughtlessly across sensitive moss beds, the areas can take years to recover. When they pilfer artifacts from the huts of early explorers, or take fossils as souvenirs from the rocky beaches, they are lost forever. And when excited tourists push their way into a colony of penguins for a "selfie," they risk destroying the unhatched eggs and disturbing the natural behavior of the penguins that scientists are anxious to study.

On a far more serious level, tourist ships have occasionally come to grief and sunk, spewing oil into the freezing water where it can cover the coats of the penguins and seals before being ingested by the adults and their young as well as affecting marine life on the sea floor. The funnels of the tourist ships also pump their exhaust into the otherwise unpolluted Antarctic air. In the past, rubbish and sewage from tourist ships has been dumped into the sea and the broken equipment of adventurers has been left abandoned on the ice. Apart from the environmental damage, the unintended effects of the tourist industry can frustrate the work of those scientists who are attempting to study the Antarctic free of the effects of human impact. On the other hand, many tourists become passionate advocates for Antarctica and can help shape public opinion in ways that promote the protection of the continent.

Unregulated, there is little doubt that the booming tourist industry would have caused even greater damage to the environment and more disruption to scientific programs. However, the treaty signatories have introduced an increasingly restrictive code of conduct for tourist operators, with the industry embracing the restrictions as a way of ensuring their continued access. Despite the restrictions, it is inevitable that the increasing number of tourists will have a negative effect on the ecology of the places they visit and that more tourist ships will run aground or sink in Antarctica's isolated and dangerous waters.

While the tourist ships are mostly restricted to the waters off the Antarctic Peninsula, adventure tourists often head for other parts of the continent. The activities of these adventurers are more difficult to control. Some of them refuse to recognize the existence of any authority in Antarctica while still expecting to be rescued if they come to grief. They continue to be a burden on scientists if they arrive at their bases, sometimes unannounced, and expect to find food and shelter, or if they call for help after suffering a mishap. When these adventurers fail to operate under the guidelines established by the Antarctic Treaty signatories and the Antarctic tourist organization, they can have a damaging impact on the areas they visit. So many climbers have now ascended Mount Vinson that their abandoned equipment and stores must surely litter its heights, in the same way that climbers have made Mount Everest into a rubbish dump. At the moment, this can't be prevented or punished, since no countries exercise sovereignty over Antarctica and no sanctions can be imposed when individuals or groups despoil the environment.

*In what ways has the popular view of Antarctica changed
in recent times?*

For centuries, Antarctica was a place of mystery and speculation. Amundsen and Scott laid much of that speculation to rest
and established Antarctica as a place for derring-do, where the
manhood of nations could play out their rivalries in the most
isolated place on Earth and gain plaudits for themselves and
prestige for their nations. By sweeping away some of its mystery, the explorers' photographs and films brought the continent closer, while also confirming its ethereal nature. Yet the
feats of those early explorers showed that the continent was
capable of being conquered. That conquest accelerated when
aviators began flying across Antarctica, establishing the equivalent of frontier towns on its coast and returning with even
more spectacular photographs and films.

While the public remains fascinated by Antarctica, some
have also begun to be concerned about the toll being taken by
the intrusion of man. This is not entirely new. A popular outcry
against the harvesting of penguins on Macquarie Island led to
a ban on their killing in 1919. Whales were not so fortunate.
It was only the realization that they were being driven to extinction that led to the establishment of the IWC. Even then,
Antarctica continued to be regarded as a place for humans to exploit, albeit now in a sustainable way, although the difficulties
of doing so ensured that the continent itself remained largely
untouched. That didn't stop suggestions after World War II for
the establishment of settlements in Antarctica and for its use as
a sanatorium or a deep freeze for storage of the world's food
surpluses. There were even suggestions that atomic bombs
could clear Antarctica of its ice.

It was not until the rise of the environmental movement in
the 1960s and 1970s that there were calls for Antarctica to be
kept in a pristine state. Wildlife films by people such as David
Attenborough, and even the popular 2007 children's animated
film *Happy Feet*, helped to bring the beauty and fragility of

Antarctica's ecosystem to the world's attention, while the oil industry's despoliation of the Arctic made environmentalists determined to prevent a similar outcome in the Antarctic. Of course, the ethereal quality of Antarctica continues to be a great attraction for tourists and adventurers. Each successful feat, and each cruise by tourist ships, helps diminish the isolation and dangers of Antarctica. When groups of women ski their way to the South Pole, as two British women did in 2000, or four Palestinians and four Israelis combine to climb an Antarctic mountain in 2004, naming it the Mountain of Israeli-Palestinian Friendship, they normalize and almost civilize a place that still tests the courage, endurance, and ingenuity of those who venture there.

Antarctica has come to be regarded as a place reserved for the pursuit of science. As the world comes to terms with the rapid pace of climate change, Antarctica is the place that holds the key to future life on Earth. With its ice sheet holding 90 percent of the world's fresh water, the prospect of it melting holds terrors for us all. If a sizable part of it melted, it would submerge some of the largest cities, from London and St. Petersburg to Shanghai, Tokyo, and New York. Low-lying countries like Holland and Denmark could also be swamped. The degree of global warming that would be required to cause such an outcome would also cause large areas of the world to become too hot to be habitable.

Does anything about Antarctica remain to be discovered?

Plenty remains undiscovered in Antarctica. Until the latter half of the twentieth century, explorers concentrated on the geology, biology, and geography of Antarctica. They collected rocks and fossils, filled specimen jars with samples of marine life, and stuffed penguins, seabirds, and seals. Recently, this early work has taken on fresh importance, since some of these early specimens provide valuable baselines for scientists seeking to make comparisons with present-day specimens.

The impact of ocean acidification on the exoskeletons of krill is made easier by museums having specimens of krill that were netted by explorers a century or more earlier. The comparisons help to buttress the observations that scientists have made in their laboratory tanks far from Antarctica, where krill have been subjected to seawater with varying levels of acidity and increasing levels of carbon dioxide. These observations and experiments have helped in understanding how krill will react to global warming if it can't be stemmed and then reversed. There remains much for scientists to learn about the life cycle of krill and their place in the food chain, and how each part of that chain will be influenced by the warming of the planet.

If acidification causes krill to decline and even disappear, there will be dire effects on all the larger life forms that rely upon krill for their food. While krill will be impacted adversely by global warming, some other creatures in the Southern Ocean are also under great threat. The species of phytoplankton upon which krill depend for food are less than a pinhead in size. Yet they have helped to shape the ways that greenhouse gases have affected life in the Southern Ocean. There is much more scientific work to be done to understand these processes before the long-term implications for the complex ecology of the Southern Ocean are clear. It will depend on how readily these life forms and the creatures that feed upon them can adapt to the warmer water, to the greater concentration of carbon dioxide in the ocean depths, and to the increased acidification of the water. Unfortunately, the fact that the acidification is occurring faster than at any time in the past 300 million years does not provide much hope for a good outcome.

The scientific research into these small life forms is not just about improving our understanding of global warming. An increasing amount of research is searching for ways to profit from the attributes of Antarctica's plant and animal life. This "bioprospecting" promises rich rewards for those who can identify a new form of biological material that could be of economic use to the world. This research can also bring new drugs to treat diseases. Because Antarctic life forms are adapted to extreme climatic conditions, they often exhibit characteristics that cannot be found in life forms located elsewhere. For instance, grass that is found on the northern tip of the Antarctic Peninsula might be used to develop frost resistance in other grasses and crops, while Antarctic bacteria that survive the freezing conditions by producing polyunsaturated fatty acids and cold-active enzymes might be developed for use by the food industry. Identifying and collecting samples of such bacteria, along with unique varieties of fungi and algae, and then artificially synthesizing the desired ingredient in the laboratory could contribute economic benefits much greater than those that the whaling industry provided in the past. More importantly, it can usually be done with little or no impact on the environment.

Because the restrictive clauses of CCAMLR are only concerned with larger life forms, there is no legal impediment or regulation on how such bioprospecting is done and who owns the results of the work. And there is no restriction on keeping the results of the bioprospecting secret until they have been patented, despite the Antarctic Treaty stipulating that scientific results are to be disclosed to all the signatories. In fact, many such patents have already been issued in Europe and the United States. Although the Antarctic Treaty nations have been divided on how to regulate bioprospecting, they will need to do so, either by adding clauses to the Antarctic Treaty or to CCAMLR or by agreeing on a new convention that is devoted just to bioprospecting.

From the smallest bacteria to the largest whales, there is so much more to be learned about life in Antarctica. Until relatively recently, scientific studies were done mostly on the assumption that these creatures were continuing to enjoy an unchanging existence in a habitat that had remained largely in the same state for millennia. Of course, that wasn't true for overexploited whales and seals, but it was mostly true for the penguins, birds, and fish. That sense of studying a static ecosystem can no longer be sustained. Instead, there is a sense of urgency about recent scientific studies that are examining how Antarctica's major life forms are being affected by climate change and how they are responding to the challenge. There is much more to know, and it is imperative that the world knows the answers to these questions, because the damage being done to the Antarctic ecosystem will affect the natural world in places far distant from the Southern Ocean.

The study of Antarctic glaciers used to be the subject of academic tomes that would collect dust on library shelves. It was one of the interests of the Australian explorer Douglas Mawson in the early part of the twentieth century, but mainly because of what the glacial moraines could tell him about the geological composition and origins of the continent. Now it is the glaciers themselves that have become of utmost interest to scientists as they try to understand how glaciers drain the ice sheets of West and East Antarctica as the supporting ice shelves break away. To this end, it is vital that we know more about how the ocean currents impact the various levels of the ice shelves, eating away at their edges and eroding them from beneath until the shelves suddenly start to collapse. Understanding how the process is playing out around the Antarctic coastline is crucial for scientists to predict accurately the rate at which sea levels are going to rise. Will New York and other coastal cities be under water in three hundred years or three thousand years, or even in a shorter timeframe altogether? Scientists are divided as to whether the process in West Antarctica is already unstoppable or whether the stemming of greenhouse gas emissions could avert the looming catastrophe.

By drilling cores of ice from depths of more than a kilometer, scientists can examine previous warming periods and the pace and severity of the intervening ice ages. Antarctica's massive ice sheet holds the key to predicting the pace of global warming and its effects. The snow that fell to create the present ice sheet contains bubbles of air from which scientists can calculate the past concentrations of oxygen, carbon dioxide, and methane. Extrapolating from this data, they can deduce the climates that prevailed over the last tens of thousands or even millions of years. The thicker the ice the better, because it can tell a story that stretches further into the past. A 3-kilometer-thick part of the ice sheet can provide evidence of the world's climate over the last 1.5 million years.

Several cores have already been taken from lesser depths, and they have provided startling details about past instances of global warming and worrisome comparisons with the global warming that has been gathering pace over the last century and a half. One recent scientific study of the gases trapped in an ice core spanning 54,000 years found that plants around the world have been responding to the increasing concentration of carbon dioxide by growing faster over the last century than at any other time. Along with the oceans, which absorb about a quarter of the fossil fuel emissions, the plants have been absorbing another quarter of the emissions that have been produced since the industrial revolution, with the remarkable plant growth keeping the world less hot than it might otherwise have become. While that is somewhat reassuring, and this information can make current climate models more accurate, neither plants nor oceans can continue mitigating the worsening effects of global warming if greenhouse gases keep increasing at the present rate.

Similar insights from even earlier periods can be found by drilling deep below the seabed to look for clues in the layers of sediment that have been deposited on the ocean floor for millions of years. These studies might explain how the Pliocene period of 3–4 million years ago experienced a similar level of carbon dioxide to the one being experienced today, yet it had temperatures that were 2–3 degrees warmer and the sea level was about 25 meters higher. As more knowledge is gathered from the Antarctic, and better computer models are developed, scientists should be able to make increasingly confident predictions about how global warming is going to affect

Antarctica and the Southern Ocean, and how that will then affect the world.

Will the Antarctic remain in a relatively pristine state?

No, of course not. To the casual observer, Antarctica appears to be an unchanging environment, frozen in its purity. Perhaps it's the influence of all that untrodden snow that makes us think of the continent as being untouched and unspoiled. While humans have only been visiting Antarctica for about two centuries, the atmospheric effects of their presence in other parts of the world can be detected in the ice that has built up from earlier centuries. When they did start stepping onto the ice, humans have left their mark on the places they have visited and sometimes inhabited. They have killed and contaminated wildlife, they have abandoned stores and equipment, they have discharged their waste into the sea, and they have buried radioactive waste in the ice. Ships have polluted Antarctica's waters when they have sunk off its coastline, and aircraft and helicopters have crashed onto its ice and been left where they fell. Entire bases have been abandoned and their detritus has been scattered by the ferocious winds. Scientists have contaminated the environment with radioactive isotopes for their experiments and drilled their way down to lakes that have existed untouched beneath the ice cap for millennia. If a map could be drawn of human activity in Antarctica and its surrounding waters over the last two hundred years, few parts of the map would be blank and untouched. While Antarctica can't truthfully be described any longer as being pristine, its sheer size has ensured that the continent, along with its atmosphere and seas, remained relatively unspoiled in comparison to the rest of our world.

The 1959 Antarctic Treaty has been fairly effective in restricting activities that would despoil the environment or decimate the wildlife. However, its lack of enforcement ability means that nations can ignore the decisions and recommendations of the treaty signatories. While some nations have made a good effort to clean up their bases, others continue to discharge waste into the ocean and in other ways neglect their responsibility to protect the environment. Rusting equipment and drums of oil can still be found around the bases of nations that shrink from the expense and effort of restoring their sites. It's debatable whether the Antarctic Treaty system is currently capable of ensuring that the protection of the Antarctic environment is taken seriously by all the nations that have bases there. Declaring Antarctica as a world park and placing it under an international governance regime that has environmental protection as its paramount objective could see Antarctica better protected. But it is the activities of humans far removed from Antarctica that are having the greatest impact on its environment.

Will the Antarctic Treaty and its associated agreements continue to govern human activity in Antarctica?

The Antarctic Treaty has been in place for nearly sixty years. In view of the passion with which some countries prosecuted their territorial claims, it was an extraordinary diplomatic triumph when they agreed to set their claims to one side for the greater good and ensure that Antarctica would henceforth be restricted to peaceful scientific endeavor. The clauses of the treaty were drawn deliberately to minimize the possibility of conflict, not only between its signatories but also with other nations that might aspire to have a presence in Antarctica, as well as with nongovernmental parties, such as scientists and environmentalists who have been allowed to become part of the treaty system.

As a result, everyone seems to share an interest in having the treaty remain beyond serious challenge. Although territorial aspirations continue to be harbored in varying degrees by many of the signatories, and continue to be advanced in surreptitious ways, the former competition for territory has been largely replaced by scientific cooperation that reaches across the imagined territorial boundaries. Despite the success of the treaty, there will continue to be debate about which parties should have the most say in the governance of the continent, whether it be the diplomats, the scientists, or the environmentalists. On some issues, the views of scientists align with those of the environmentalists, but they also have their own agendas and can regard environmental restrictions on their research activities as irksome. They don't want to be prevented from using radioactive isotopes out in the field or drilling an ice core down to a previously unsullied lake beneath the ice cap. Yet both sides have managed so far to compromise and coexist.

The nations with territorial claims have also accepted the presence of many other nations in Antarctica and have mostly managed to reach a peaceful agreement on issues that could have proved problematic. The banning of mining under the 1992 Madrid Protocol was one such issue that could have severely tested the treaty system, but the treaty's inherent flexibility and its value to the signatories ensured that the dispute could be peacefully resolved without causing a ruinous split in the organization. Indeed, it is likely that the treaty will continue to be free of serious challenge for the foreseeable future, with the signatories and other interested parties muddling along in the spirit of compromise and cooperation that has characterized the governance of Antarctica for more than half a century. This is fortunate, because it is difficult to see what sort of widely respected agreement could take its place.

Can we stop Antarctica from melting?

The scientific consensus about the melting of Antarctica has changed radically in recent years. Previously, the measurements of sea ice, the amount of snowfall, the protective function of the cold circumpolar current, and the apparently unchanging ice shelves combined to suggest that Antarctica might be shielded from the worst effects of global warming and even be able to offset the dramatic changes being witnessed in the Arctic. That confidence gradually disappeared as scientists gained a better understanding of the ice shelves and the glaciers that feed them, as well as the workings of the ocean currents in the Southern Ocean. It is now clear that increased snowfall will be insufficient to outweigh the increased ice loss as the collapsing ice shelves cause an acceleration in the movement of many of the glaciers and a consequent thinning of the ice sheet.

As a result, the former equanimity among some scientists and policymakers has been replaced by a growing conviction that serious melting and rising sea levels will occur during the lifetimes of many people now living. The latest scientific modeling suggests there will be a rise of nearly 3 meters (about 10 feet) in sea levels by 2100, which will have catastrophic effects on low-lying coastal areas and many crowded cities. Moreover, there will be increasingly serious impacts on the world in the decades leading up to the end of this century. The big question facing humanity is whether a change in human behavior could slow down or even reverse this process.

The answer to this crucial question may be different for East and West Antarctica. The quick collapse of some ice shelves in West Antarctica, and the acceleration of their glaciers, has caused glaciologists to suggest that the situation in West Antarctica may have reached a tipping point from which it will be difficult, if not impossible, to recover. Indeed, the scientific records from West Antarctica have revealed that since 1950 it has experienced a temperature increase of about 0.5°C per decade, making it one of the world's fastest-warming places. The warmer air and water has been eating away at its ice shelves, which has recently caused some of them to suddenly collapse. On land, its glaciers are retreating and there is a growing area that is clear of snow and ice.

Scientists now understand that the ice of West Antarctica is more of an ice basin rather than an ice cap, and that the loss of its ice shelves will cause an inexorable draining of that basin as huge cliffs of ice move toward the sea and break off. There would have to be a substantial reduction in the temperatures of West Antarctica to have any hope of preventing the collapse of its remaining ice shelves and stemming the movement of its ice cap toward the sea. Such a reduction could be possible over an extended time period if the world can quickly stop the discharge of greenhouse gases and contain the postindustrial temperature increase below the manageable level of 1.5°C that was agreed at the Paris climate conference in December 2015.

While weather systems helped to raise the postwar temperature of West Antarctica by about 3°C, the air temperature records of East Antarctica have held steady or even decreased in some places on the ice cap. This is because the East Antarctica ice cap creates its own climate, with its cold katabatic winds keeping warmer air from the Southern Ocean at bay and the circumpolar current keeping the Southern Ocean's warmer surface water away from the continent. The temperature records helped to reassure scientists that East Antarctica might not mimic the experiences in West Antarctica and Greenland. Until recently, the nature of East Antarctica and the heavier snowfalls that have been recorded on the ice cap in recent decades led scientists to believe that it was accumulating more ice on the ice cap than it was losing from the thinning ice shelves around the coastline.

While there is still much for scientists to learn about the complex relationships among the ice cap, the ice shelves, and the ocean currents of East Antarctica, it has recently become clear that East Antarctica will not be immune to the influence of global warming. The annual measurement of the surface area of the ice shelves in East Antarctica gave a misleading impression that the shelves were withstanding the effects of the warming water temperature, with the cold circumpolar current supposedly keeping the warmer water away from East Antarctica. However, recent measurements have revealed that the ice shelves are being melted from below by warming currents in the deep ocean that manage to flow beneath the circumpolar current. Meanwhile, meltwater lakes and craters have been observed on the surface of at least one ice shelf, with a waterfall sending water cascading through the crater into the ocean below. As they melt from above and are eroded from below, some of the ice shelves of East Antarctica are proving to be just as prone to collapse as the ice shelves of West Antarctica and Greenland, whereupon they will release their braking influence on the glaciers that feed them. As the glaciers speed up, they will cause a thinning of the ice cap behind them and a great dumping of water into the Southern Ocean.

The thinning of the ice cap and the discharge of water into the ocean is also occurring directly from the surface of the ice cap, as meltwater flows from deep inland to the coast. When the water pools on the surface of ice shelves, it increases their melting and hastens their slow disintegration. This phenomenon has been observed in Greenland but was not believed to occur in Antarctica. However, recent studies of aerial photographs and satellite images have revealed pools of meltwater, not only on ice shelves but on the ice cap at heights of more than a thousand meters. In the summer, rivers of meltwater can be seen flowing toward the ocean. This alarming development has to be incorporated into the new computer models before scientists can understand the full implications for the rate of the future sea-level rise. What is clear is that it adds urgency to the need to cut back on the discharge of greenhouse gases, until the world is no longer exacerbating the pace of global warming.

Will the Antarctic ice cap ever melt completely?

As to the question of whether the Antarctic ice cap will ever melt completely, the answer is perhaps. The Antarctic ice cap is up to 4 kilometers (2.5 miles) deep and contains sufficient water to cause the sea level to rise by a catastrophic 58 meters (190 feet) if it was all to melt. Only about 10 percent of that rise would come from West Antarctica and the rest from East

Antarctica. That being said, the Antarctic ice cap would take a long time to melt, if it ever does. With the limits of their knowledge about the complex interactions between the Antarctic atmosphere and the Southern Ocean, along with the ice shelves and glaciers, and with the consequent limits of their computer models, scientists can't say with any certainty how long it will take. Predictions depend upon a multitude of factors that are out of the control of humans. But some factors are within our control, particularly our greenhouse gas emissions.

If nothing is done about the burning of fossil fuels and the discharge of greenhouse gases, the timeframe for the melting will be much shorter than if the Earth's atmosphere was restored to the levels that prevailed a century ago. While initial estimates suggested that it could take up to several thousand years for Antarctica to become free of ice, more recent estimates have reduced the timeframe to as little as five hundred years. Well before that happens, there will be an increasingly calamitous rise in sea levels, reaching nearly 3 meters (10 feet) by the end of this century. Despite such dire possibilities, the much greater volume of ice in East Antarctica, combined with its intensely cold climate and protective winds, offers some chance of stemming the sea level rise before it overwhelms us. If nations can achieve a carbon-neutral world, or better still a carbon-negative world, they might yet slow the melting of East Antarctica and, over several centuries, even restore the continent to its former state.

NOTES

Chapter 1

1. David Day, *Antarctica: A Biography* (New York: Oxford University Press, 2013), 22.
2. Edouard Stackpole, *The Voyage of the Huron and the Huntress: The American Sealers and the Discovery of the Continent of Antarctica* (Mystic, CT: Marine Historical Association, 1955), 51.
3. Christine Holmes (ed.), *Captain Cook's Second Voyage: The Journals of Lieutenants Elliott and Pickersgill* (Dover, NH: Caliban Books, 1984), 40–41.
4. William Stanton, *The Great United States Exploring Expedition of 1838–1842* (Berkeley: University of California Press, 1975), 180–85.

Chapter 2

1. C. E. Borchgrevink, *First on the Antarctic Continent* (London: George Newnes, 1901), 84; Louis Bernacchi, *To the South Polar Regions* (London: Hurst and Blackett, 1901), ix–x.
2. Karl Fricker, *The Antarctic Regions* (London: Swan Sonnenschein, 1900), 278–80.
3. Roland Huntford, *Scott and Amundsen* (London: Hodder and Stoughton, 1979), 282–83.
4. Huntford, *Scott and Amundsen*, 515–17.
5. Shirase Expedition Supporters Association, *The Japanese South Polar Expedition 1910–12* (Norwich: Erskine Press, 2012), 15.

Chapter 4

1. David Day, *Antarctica: A Biography* (New York: Oxford University Press, 2013), 396.

INDEX